MESSLEY

pg 67

Ways to Teach Biology

American University Studies

Series XIV
Education

Vol. 23

PETER LANG
New York • Bern • Frankfurt am Main • Paris

Sharon La Bonde Hanks

Ways to Teach Biology

The Whys and Hows of Changing to a Process Approach

PETER LANG
New York • Bern • Frankfurt am Main • Paris

Library of Congress Cataloging-in-Publication Data

Hanks, Sharon La Bonde
　　Ways to teach biology : the whys and hows of
changing to a process approach / Sharon La Bonde Hanks
　　　　p. cm. — (American university studies. Series XIV,
Education ; vol. 23)
　　　　Bibliography: p.
　　　　1. Biology—Study and teaching.　I. Title.
II. Series.
QH315.H274　　　1989　　　574.07—dc19　　　89-2271
ISBN 0-8204-0957-X　　　　　　　　　　　　CIP
ISSN 0740-4565

CIP-Titelaufnahme der Deutschen Bibliothek

LaBonde Hanks, Sharon:
Ways to teach biology : the whys and hows of
changing to a process approach / Sharon LaBon-
de Hanks. — New York; Bern; Frankfurt am
Main; Paris: Lang, 1989.
　(American University Studies: Ser. 14,
　Education; Vol. 23)
　ISBN 0-8204-0957-X

NE: American University Studies / 14

© Peter Lang Publishing, Inc., New York 1989

Printed by Weihert-Druck GmbH, Darmstadt, West Germany

TO MY FAMILY AND FRIENDS
FOR THEIR CONTINUOUS SUPPORT AND ENCOURAGEMENT

TABLE OF CONTENTS

CHAPTER I

INTRODUCTION

OVER THE LAST FEW YEARS, MY IDEAS ABOUT WHAT TEACHING IS AND WHAT I'M SUPPOSED TO BE DOING OR WANT TO DO IN CLASS HAVE CHANGED. I AM VERY EXCITED ABOUT THE CHANGES THAT HAVE OCCURRED AND THEREFORE HAVE UNDERTAKEN THIS PROJECT: TO EXPLAIN HOW I TEACH GENERAL BIOLOGY I AND II AND WHY I TEACH THE WAY I DO. THIS IS DIFFICULT BECAUSE THERE ARE SO MANY DIFFERENT IDEAS, CONCEPTS, AND STRATEGIES THAT NEED TO BE SORTED OUT. I HAVE COME TO REALIZE THAT UNLIKE MOST OF MY PROJECTS, WHICH ARE LINEAR (A-B-C), THIS ONE IS MORE LIKE A JIG SAW PUZZLE. SOME PIECES (IDEAS) ARE BRIGHTLY COLORED AND SEEM AT FIRST TO BE MORE ESSENTIAL TO THE PUZZLE AS THEY FORM THE PICTURE WHILE OTHER PIECES CONTRIBUTE TO THE BACKGROUND AND THEREFORE SEEM UNIMPORTANT. HOWEVER, WITHOUT BACKGROUND PIECES THE PUZZLE WOULDN'T HOLD TOGETHER. EACH PIECE FITS INTO TWO OR THREE OTHER PIECES WITH ALL CONTRIBUTING TO THE FINAL PUZZLE. SINCE I CAN'T START WITH A AND THEN PROGRESS TO B AND THEN TO C IN A LINEAR FASHION AND I CAN'T TALK JUST ABOUT THE COLORED PIECES OF THE PICTURE, I WILL JUST HAVE TO PICK A PLACE TO START AND BY THE TIME THE PUZZLE IS FINISHED I HOPE THAT THE WHOLE PROCESS, STRUCTURE, FUNCTION AND INTERRELATIONSHIPS OF THE PIECES ARE APPARENT.

AS I LOOK BACK OVER THE LAST FEW YEARS THE PROCESS OF TRANSFORMATION OR CHANGE STARTED BECAUSE I LIKE TO LEARN ABOUT NEW IDEAS OR DO NEW PUZZLES. I'D PICK UP DIFFERENT PIECES, LOOK AT THEM (COLOR, SHAPE, CONTENT) AND PUT THEM DOWN. SOME PIECES SEEMED TO FIT TOGETHER SO I ASSEMBLED THEM INTO SMALL SECTIONS. AFTER ENOUGH SMALL SECTIONS WERE ACCUMULATED I WOULD SUDDENLY SEE HOW THEY COULD FIT TOGETHER, MAKING EVEN LARGER SECTIONS. I WORKED AT THE PUZZLE OFF AND ON AS MY INTEREST AND TIME ALLOWED. IN THE BEGINNING I HAD NO BURNING DESIRE OR DRIVE TO FINISH THE PUZZLE. HOWEVER, A DAY FINALLY CAME WHEN I THOUGHT I COULD SEE THE PICTURE IN THE PUZZLE AND THEN ALL I WANTED TO DO WAS PUT-IT-ALL-TOGETHER, TO ARRANGE ALL THE SEPARATE SECTIONS SO THEY MADE SENSE, TO FINISH THE PUZZLE, TO HAVE THE WHOLE IN FRONT OF ME. THE ONLY DIFFERENCE IS THAT EVEN THOUGH RIGHT NOW THE PUZZLE SEEMS FINISHED I KNOW THAT THE PICTURE WILL SHIFT AND CHANGE AS I FIND NEW PIECES NOT YET IN MY PUZZLE BOX. SO WHAT IS PRESENTED HERE IS THE PUZZLE TO DATE. THIS IS MY ATTEMPT TO EXPLAIN WHAT HAS HAPPENED WITH MY TEACHING OVER THE LAST TEN YEARS.

I MUST ADMIT THAT I NEVER PLANNED TO BECOME A TEACHER. I DECIDED TO GO TO GRADUATE SCHOOL IN 1965 BECAUSE I WANTED TO LEARN MORE ABOUT BOTANY. I LOVED TO LEARN NEW THINGS AND FOUND AS A GRADUATE TEACHING ASSISTANT IN A GENERAL BIOLOGY COURSE FOR UNDERGRADUATES, THAT I LIKED TEACHING NEW IDEAS, CONCEPTS AND THEORIES TO OTHERS. BECAUSE I NEVER PLANNED ON BECOMING A TEACHER I HAVE NEVER TAKEN ANY EDUCATION COURSES. I TAUGHT THE

WAY I LIKED TO LEARN OR AS I SAW IT THEN AS THE WAY GOOD TEACHERS, THE ONES I LIKED, TAUGHT. I PREFERRED A HIGHLY ORGANIZED LECTURE FORMAT PRESENTED IN A STRAIGHT FORWARD MANNER. UP UNTIL A FEW YEARS AGO MY TEACHING STYLE WAS WHAT WHITE REFERS TO AS "FILLING THE BUCKET", I LECTURED OR POURED CONTENT INTO THE STUDENTS' HEADS. OVER THE YEARS I HAVE TAUGHT A VARIETY OF COURSES: UNDERGRADUATE COURSES FOR MAJORS AND NON-MAJORS, GRADUATE COURSES FOR MASTERS CANDIDATES, LECTURE-LABORATORY COURSES, SEMINARS, GENERAL COURSES LIKE GENERAL BIOLOGY AND SPECIALIZED COURSES LIKE SCANNING ELECTRON MICROSCOPY OR ECOLOGY. I HAVE CHOSEN TO USE, AS MY EXAMPLE FOR THIS PROJECT THE COURSE I TEACH MOST FREQUENTLY, THE YEAR LONG SEQUENCE OF GENERAL BIOLOGY I AND II OFFERED FOR SCIENCE MAJORS.

IF A STUDENT WHO HAD TAKEN GENERAL BIOLOGY WITH ME THREE YEARS AGO WERE TO VISIT MY CLASS TODAY, THE CONTENT WOULD CERTAINLY BE FAMILIAR BUT WHAT GOES ON DURING THE CLASS WOULD BE DIFFERENT. I HAVE GONE FROM A STRAIGHT LECTURE-QUESTION APPROACH TO TEACHING IN A CONTENT-PROCESS ORIENTATION. LIKE MANY CHANGES IN LIFE I DID NOT START OUT BY SAYING "I'M GOING TO CHANGE, THIS IS MY GOAL AND I WILL WORK TOWARDS IT ONE STEP AT A TIME." IN FACT, IF SOMEONE HAD ASKED ME, I WOULD HAVE SAID THAT I DIDN'T WANT TO CHANGE AT ALL OR EVEN SEE THE NEED, I LIKED HOW I DID THINGS, I WAS COMFORTABLE WITH MY TEACHING, MY STYLE WAS ACCEPTABLE TO MY PEERS AND FIT WITH MY OWN LEARNING STYLE (WHICH I DIDN'T KNOW AT THE TIME). CHANGES, ON THE OTHER HAND ARE RISKY, MAKE ME FEEL UNCOMFORTABLE, INSECURE, SELF-DOUBTING AND REQUIRE A LOT OF MENTAL EFFORT AND WORK. THUS, CHANGE HAS CREPT UP ON ME AS A SERIES OF SMALL SEPARATE QUANTITATIVE EXPERIENCES (MANY UNLOOKED FOR) WHICH HAVE RESULTED IN A MAJOR QUALITATIVE CHANGE IN HOW I THINK AND HOW I TEACH. THE CHANGE WHICH OCCURRED IN FITS AND STARTS OVER THE PAST TEN YEARS HAS ACCELERATED IN THE PAST TWO YEARS. PROGRESSING FROM SMALL CHANGES (ONE HERE AND ONE THERE) UNTIL A CRITICAL POINT WAS REACHED WHEN COMMITMENT TO THE IDEAS OF PROCESS AND WRITING FORCED A COMPLETE REVOLUTION IN WHAT I DO AND HOW I DO IT.

THE REALLY GREAT THING ABOUT CHANGE IS THAT IT IS NEVER FINISHED, THERE IS ALWAYS SOMETHING NEW AND EXCITING TO LEARN AND TO FIT INTO YOUR EXPERIENCE PERSONALLY AND AS A TEACHER.

WHY HAVE I CHANGED HOW I TEACH? I WAS COMFORTABLE AND I THINK SO WERE THE MAJORITY OF MY STUDENTS WITH "FILL THE BUCKET" APPROACH. WHY CHANGE TO A DIFFERENT TEACHING STYLE WHICH, BECAUSE I'M NEW AT IT AND AM STILL WORKING TO GET IT DOWN RIGHT, MAKES ME UNCOMFORTABLE WITH MYSELF AND WITH MY COLLEAGUES AND A STYLE THAT MAKES THE STUDENTS LESS COMFORTABLE WITH ME AND WITH THEMSELVES? HOW HAS THIS HAPPENED? WHAT HAS INFLUENCED ME? IN LOOKING BACK TRYING TO FIGURE OUT WHAT HAPPENED, I LOOKED FOR WHAT AND/OR WHO CONTRIBUTED TO THE CHANGE, WHAT WAS THE CHRONOLOGY, HOW DID I INCORPORATE THE EXPERIENCES AND IDEAS, AND WHAT HAS BEEN THE RESULT TO DATE. I AM GOING TO RECOUNT MY

JOURNEY FOR YOU (EVENTS, PEOPLE AND IDEAS), SHOW YOU THE CURRENT PRODUCT (OR WHERE GENERAL BIOLOGY I AND II ARE NOW), AND FINALLY PRESENT YOU WITH AN ANOTHER APPROACH TO TEACHING LABORATORIES (AS YET UNTRIED).

JUST LIKE IN ECOLOGY, THE APPROACH TO THIS PROJECT IS HOLISTIC WHERE EVERYTHING IS INTERRELATED WITH EVERYTHING ELSE. I CAN ONLY TALK ABOUT ONE THING AT A TIME AND THEN TRY TO PULL IT ALL TOGETHER IN THE GENERAL BIOLOGY I AND II SYLLABI AND ASSIGNMENTS. BELOW IS A HIGHLY SIMPLIFIED CLUSTER DIAGRAM OF THIS PROJECT. EACH OF THE LARGE CIRCLES INFLUENCES OR HAS INTERRELATIONSHIPS WITH ALL THE OTHER CIRCLES EITHER DIRECTLY OR INDIRECTLY AND EACH ITEM IN A CIRCLE MAY HAVE ONE OR MORE SUBDIVISIONS. I HAVE INTENTIONALLY NOT DRAWN IN THE ARROWS TO INDICATE ALL POSSIBLE INTERCONNECTIONS NOR HAVE I LISTED ALL THE SUBDIVISIONS AS THESE WOULD MAKE THE DIAGRAM SO CONFUSING THAT ITS USEFULNESS AS A VISUAL ORIENTATION WOULD BE LOST.

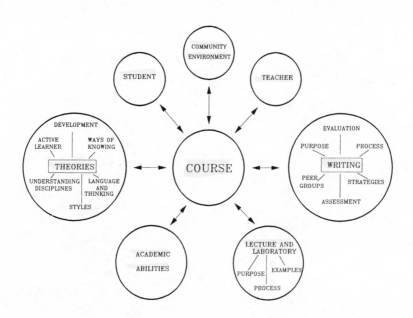

CHAPTER 2

EXPERIENCES AND CHANGE

ALTHOUGH EXCITING, CHANGE IS DIFFICULT AND DOES NOT HAPPEN IN A VACUUM. THE FOLLOWING PROGRAMS, PEOPLE AND EVENTS ALL CONTRIBUTED TO SET THE STAGE FOR MY GROWTH AND DEVELOPMENT.

FRESHMAN SEMINAR

IN 1977 I WAS ASKED BY CLIFFORD ADELMAN TO PARTICIPATE IN THE GENERAL EDUCATION STUDIES PROJECT FUNDED BY FIPSE (FUND FOR THE IMPROVEMENT OF POST SECONDARY EDUCATION). THE PROJECT RESULTED IN A COURSE CALLED FRESHMAN SEMINAR WHICH WAS TAUGHT IN A VARIETY OF TEACHING MODES (NONE OF THEM FILL THE BUCKET) USING DIFFERENT TEACHING STRATEGIES. FACULTY FROM SEVERAL DISCIPLINES SPENT A LOT OF TIME DISCUSSING HOW STUDENTS LEARN AND HOW TO HELP THEM TO BE ACTIVE LEARNERS. I WAS INTRODUCED TO THE CONCEPTS OF STUDENT OWNERSHIP, LEARNING CONTRACTS, ACADEMIC COMPETENCIES AND LEARNING STYLES. DURING THIS EXPERIENCE I TRIED TO INCORPORATE SOME OF THE TEACHING TECHNIQUES (GROUP DISCUSSION) AND LEARNING STRATEGIES (STUDENT OWNERSHIP AND CONTRACTS) FROM THE PROJECT INTO MY OTHER COURSES. BUT I FOUND THAT THEY DIDN'T WORK OR I DIDN'T KNOW HOW TO MAKE THEM WORK IN ALL MY COURSES AND THAT THEY WORKED BEST IN MY JUNIOR AND SENIOR SEMINARS. IN GENERAL BIOLOGY I AND II THERE WAS SOME DISCUSSION DURING THE FIRST CLASS SESSION ON PREFERRED LEARNING STYLE OR HOW EACH STUDENT LEARNED BEST OR WHICH WAY OF LEARNING WAS THE MOST COMFORTABLE FOR THEM. BUT GENERAL BIOLOGY I AND II WERE PRIMARILY "FILL THE BUCKET".

ENVIRONMENTAL STUDIES PROGRAM

IN 1979 CHARLES LEE WROTE AND WAS AWARDED A NSF-CAUSE GRANT (NATIONAL SCIENCE FOUNDATION - COMPREHENSIVE ASSISTANCE TO UNDERGRADUATE SCIENCE EDUCATION): FOR THE REVITALIZATION OF AN INTERDISCIPLINARY ENVIRONMENTAL STUDIES PROGRAM AND I WAS INVITED TO JOIN AS A TEAM MEMBER. THE TEAM WAS COMPOSED OF FACULTY FROM DIFFERENT DISCIPLINES WHO WERE TO DEVELOP A PROGRAM WHICH WOULD HAVE A CORE OF INTERDISCIPLINARY TEAM TAUGHT COURSES FOR FRESHMAN THROUGH SENIORS. WHILE WORKING CLOSELY WITH FACULTY FROM OTHER DISCIPLINES I LEARNED HOW THEY TAUGHT THEIR COURSES, WE BEGAN TO DISCUSS HOW TO MAKE INTERCONNECTIONS BETWEEN DISCIPLINES AND WHAT WE WANTED THE STUDENT TO BE ABLE TO DO BESIDES KNOW FACTS (CONTENT). ALTHOUGH, I WAS GAINING INSIGHT INTO OTHER APPROACHES TO TEACHING, THERE WERE STILL NO MAJOR CHANGES IN GENERAL BIOLOGY I AND II.

GENERAL EDUCATION REFORM

WORKING IN BOTH FRESHMAN SEMINAR AND ENVIRONMENTAL STUDIES PREPARED ME, IN TERMS OF INTEREST AND COMMITMENT, FOR THE UNDERGRADUATE CURRICULUM REFORM MOVEMENT WHICH HIT MY CAMPUS IN

1979 AND CONTINUES TODAY. GENERAL EDUCATION REFORM RESULTED IN LONG DISCUSSIONS OF ACADEMIC SKILLS, COMPETENCES, ABILITIES AND/OR CONTENT. ALTHOUGH THERE WAS NO SURFACE CHANGE IN GENERAL BIOLOGY, MY ORIENTATION AWAY FROM JUST CONTENT PRESENTATION HAD DEFINITELY BEGUN. THE QUESTION WAS HOW TO CHANGE AWAY FROM THE "FILL THE BUCKET" APPROACH TO TEACHING IN ORDER TO TEACH ACADEMIC ABILITIES.

WRITING AS PROCESS/WRITING ACROSS THE CURRICULUM

IN THE SPRING OF 1984, I PARTICIPATED IN A SEMINAR, WRITING AS PROCESS, PRESENTED BY VIRGIE GRANGER AND ROBERT KLOSS. THIS WEEK LONG EXPERIENCE INTRODUCED ME TO PROCESS TEACHING. SINCE THAT WORKSHOP I'VE BEEN ACTIVELY INVOLVED IN WRITING ACROSS THE CURRICULUM IN NUMEROUS WAYS FROM GOING TO WORKSHOPS, ROUNDTABLES AND SEMINARS TO ACTUALLY RUNNING WORKSHOPS AND ROUNDTABLES TO CONTRIBUTING TO ON WRITING WELL: A FACULTY GUIDEBOOK FOR IMPROVING STUDENT WRITING IN ALL DISCIPLINES (KLOSS, 1985). I HAVE BEEN A TEAM MEMBER IN DONNA PERRY'S GRANT FROM THE NEW JERSEY HUMANITIES COUNCIL: USING WRITING AS A MODE OF LEARNING AND I HAVE JOINED A GROUP OF FACULTY FROM SEVERAL DISCIPLINES WHO ARE INTERESTED IN PROCESS AND WHO HELP EACH OTHER FIND WAYS TO INCORPORATE PROCESS INTO THEIR CONTENT COURSES. I NOW HAD A NEW ORIENTATION TOWARDS MY TEACHING, I WANTED TO INCREASE CONTENT LEARNING BY USING THE WRITING PROCESS. I BEGAN, TO A LIMITED EXTENT, TO INCORPORATE WRITING AND PROCESS IN MY COURSES BUT I WAS STRUGGLING WITH THE PROBLEM OF CONTENT VERSUS PROCESS. PROCESS DOES REQUIRE TIME AND BECAUSE NOT AS MUCH CONTENT COULD BE COVERED, I WAS TROUBLED BY THE THOUGHT THAT I WAS "SACRIFICING" CONTENT FOR PROCESS.

HUMANITIES STUDIES SEMINAR

THEN IN 1985 STEPHEN HAHN AND JOHN PETERMAN OFFERED A YEAR LONG SEMINAR FUNDED BY A GRANT FROM THE NEW JERSEY DEPARTMENT OF HIGHER EDUCATION: HUMANISTIC STUDIES PROJECT FACULTY DEVELOPMENT. THE SEMINAR CENTERED ON HOW STUDENTS LEARN, THEORIES OF COGNITIVE DEVELOPMENT AND HOW TEACHING STYLES CAN EFFECT STUDENT LEARNING AND DEVELOPMENT. THE PAPERS THAT WE READ AND DISCUSSED AND THE SEMINAR PRESENTORS HELPED ME TO FINALLY UNDERSTAND WHY WHAT I HAD BEEN DOING OR TRYING TO DO FOR THE PAST FEW YEARS WORKED OR DIDN'T WORK. I DISCOVERED NEW APPROACHES AND WAYS TO IMPROVE AND CHANGE MY TEACHING TO REACH MY GOALS. THE SEMINAR HELPED ME TO FOCUS MY THINKING AND ALTER MY IDEAS ABOUT HOW TO INCORPORATE PROCESS TO INCREASE STUDENT DEVELOPMENT AND LEARNING IN MY CONTENT COURSES. THIS PROGRAM GAVE ME THE RATIONAL FOR INCORPORATING WRITING AS PROCESS IN MY LECTURE AND LABORATORIES AND EXPLAINED WHY FRESHMAN COURSES ARE DIFFERENT FROM SENIOR CONTENT COURSES AND WHY SEMINARS ARE DIFFERENT FROM LECTURE COURSES IN TERMS OF STUDENT LEARNING STYLES AND DEVELOPMENT.

IN THE FALL OF 1986, I MADE THE FINAL COMMITMENT TO COMBINE

PROCESS AND CONTENT TO INCREASE STUDENT LEARNING AND DEVELOPMENT AND TO TRANSFORM GENERAL BIOLOGY BY USING WRITING AND SMALL GROUP WORK (PEER GROUPS).

I WANT TO THANK ALL THOSE PEOPLE WHO WROTE, RECEIVED GRANTS, AND RAN THE PROGRAMS AND PROJECTS IN WHICH I WAS INVOLVED. I ALSO WANT TO THANK ALL THE GUEST PRESENTORS AND OTHER COLLEAGUES WHO HAVE SHARED THEIR IDEAS AND THOUGHTS WITH ME OVER THE YEARS. THE INFORMATION IN THIS PROJECT HAS COME FROM MANY DIFFERENT SOURCES: BOOKS, OFF-PRINTS, FACULTY HANDOUTS, WORKSHOPS, SEMINARS, CONVERSATIONS OVER COFFEE, AND MY OWN CLASSROOM EXPERIENCES. THIS PROJECT INCLUDES THE IDEAS AND STRATEGIES THAT I USE IN AND FOR GENERAL BIOLOGY I AND II. IT IS REALLY A BLEND OF WHAT I HAVE LEARNED AND EXPERIENCED. I HAVE INCLUDED A SELECT GROUP OF REFERENCES AS A STARTING POINT FOR THOSE WHO ARE INTERESTED IN HAVING A MORE IN DEPTH UNDERSTANDING OF THE IDEAS DISCUSSED IN THIS PROJECT. THIS LIST REFLECTS ONLY A SMALL PORTION OF THE MATERIALS AVAILABLE (SEE REFERENCES).

CHAPTER 3

THEORIES

EVEN THOUGH MOST OF THE TIME I AM GOING TO BE TALKING ABOUT PROCESS I WANT TO BE VERY CLEAR THAT GENERAL BIOLOGY I AND II ARE CONTENT COURSES TAUGHT FOR SCIENCE MAJORS. CONTENT IS AT THE CENTER OF THE COURSES AND I USE THE WRITING PROCESS AS A WAY TO LEARN THAT CONTENT AND ACHIEVE THE GOALS I HAVE FOR STUDENT DEVELOPMENT. EACH TEACHER HAS THEIR OWN TEACHING STYLE AND CAN ADAPT IT TO HELP THE STUDENTS DEVELOP. THIS PROJECT IS BASED ON MY TEACHING STYLE AND I HOPE THAT BY WRITING THIS I MAY HELP SOMEONE THINK ABOUT HOW THEY TEACH, IF ONLY MYSELF. JUST LIKE MY STUDENTS, I HAVE FOUND THAT WHEN I WRITE ABOUT SOMETHING I COME TO A NEW UNDERSTANDING OR SEE IT DIFFERENTLY. I EXPECT THAT AS I WRITE IT ALL OUT THAT THE EXPERIENCES, THEORIES AND PROCESSES IN MY HEAD WILL DEVELOP INTO A FUNCTIONAL WHOLE.

COGNITIVE DEVELOPMENT AND LEARNING THEORIES

ALL MY LIFE I HAVE CLASSIFIED PEOPLE, INCLUDING STUDENTS, INTO TWO BROAD CATEGORIES THOSE I CALLED THE WHITE AND BLACK THINKERS WHO SAW ONLY WRONG/RIGHT, GOOD/BAD OR THEM/US TO EVERY ISSUE AND THOSE I CALLED THE GRAY THINKERS WHO SAW THE OTHERS PERSON'S POINT OF VIEW OR SIDE OF THE ISSUE, WHO COULD WEIGH THE ARGUMENTS AND MAKE A DECISION EVEN IF IT MEANT CHANGING THEIR MINDS. I HAVE INTUITIVELY KNOWN FROM CHILDHOOD ABOUT THESE TWO GROUPS OF THINKERS BUT NEVER UNDERSTOOD WHY THEY WERE THE WAY THEY WERE. I HAVE ALSO BEEN ACUTELY AWARE FOR THE LAST FEW YEARS THAT THE TEACHING METHODS I USED IN MY FRESHMAN LECTURE/LABORATORY CLASSES AND THOSE I USED IN MY JUNIOR/SENIOR SEMINARS WERE VERY DIFFERENT AND THAT I COULDN'T EASILY TRANSFER THE SEMINAR METHODS INTO MY FRESHMAN CLASSES. AGAIN I DIDN'T FULLY UNDERSTAND WHY THAT WAS TRUE. THEN, I WAS INTRODUCED TO WILLIAM PERRY'S (1981) SCHEMA OF COGNITIVE DEVELOPMENT. WIDICK AND SIMPSON (1978, P.46) SAY IT BEST FOR ME, PERRY'S THEORY IS "FINDING A THEORY THAT MAKES SENSE OUT OF EXPERIENCE". SUDDENLY THINGS JUST FELL INTO PLACE. ALTHOUGH OTHER RESEARCHERS DO NOT COMPLETELY AGREE WITH THE PERRY SCHEMA, IT WORKS FOR ME SO I HAVE INCORPORATED IT ALONG WITH OTHER THEORIES INTO MY IDEAS ABOUT TEACHING.

WILLIAM PERRY'S SCHEMA, BASED ON RESEARCH INITIALLY DONE WITH MALE STUDENTS BUT LATER REPEATED WITH FEMALE STUDENTS, DEALS WITH THE COGNITIVE AND ETHICAL DEVELOPMENT OF STUDENTS. HE HAS DIVIDED THE DEVELOPMENT PROCESS INTO NINE POSITIONS OR STAGES WITH CLEARLY DEFINED TRANSITIONAL PHASES. HE DESCRIBES EACH STAGE IN TERMS OF HOW STUDENTS SEE AND RELATE TO THE WORLD, KNOWLEDGE, EDUCATION, ONESELF AND VALUES. SUMMARIZED, THE PROGRESSION IS FROM DUALISM (TWO CHOICES RIGHT/WRONG, THE AUTHORITY IS RIGHT) THROUGH MULTIPLICITY (A DIVERSITY OF OPINIONS IS OK, EVERY OPINION IS EQUALLY RIGHT) TO RELATIVISM (A DIVERSITY

OF OPINIONS IS OK BUT BASED ON THE EVIDENCE ONE CAN SELECT A SINGLE JUDGMENT) TO THE FINAL STAGE OF COMMITMENT IN RELATIVISM (A CHOICE FOR COMMITMENT BASED ON RELATIVISTIC THINKING) (PERRY, 1981, P.79-80). STUDENTS ADVANCE THROUGH THE POSITIONS OR DEVELOP AT DIFFERENT RATES. SEVERAL RESEARCHERS HAVE RELATED PERRY'S DEVELOPMENT SCHEMA TO HOW STUDENTS LEARN, HOW WE CAN TEACH TO DEVELOP STUDENT COGNITIVE GROWTH, HOW TEACHING STYLE AND STUDENT DEVELOPMENT FIT, AND HOW TO STRUCTURE COURSES OR FRAME CLASSES TO PROVIDE FOR PROGRESS IN STUDENT DEVELOPMENT. I AM GOING TO DISCUSS THE DUALIST STAGE IN SOME DETAIL AS MOST OF MY FRESHMAN STUDENTS FALL INTO THAT CLASSIFICATION. AFTER I HAVE COVERED THE CHARACTERISTICS OF A DUALIST I WILL DISCUSS WHAT CAN BE DONE TO MOVE STUDENTS TOWARD THE MULTIPLISTIC AND RELATIVISTIC POSITIONS AND FINALLY I WILL TALK ABOUT MATCHING TEACHING STYLES TO STUDENTS LEARNING STYLES (PRIORITY). THESE IDEAS ARE BASED PRIMARILY ON WORKS OF PERRY (1981), WIDICK AND SIMPSON (1978) AND MAGOLDA (1985).

THE DUALIST VIEWS ALL KNOWLEDGE AS KNOWN. THERE ARE RIGHT ANSWERS AND WRONG ANSWERS AND ALL STUDENTS HAVE TO DO IS LEARN (MEMORIZE) THE RIGHT ANSWERS. THIS IS PERCEIVED AS LEARNING. THE INFORMATION IS PRESENTED BY THE ALL KNOWING AND THE ONLY LEGITIMATE AUTHORITY - THE TEACHER AND PEER OPINIONS DO NOT COUNT. A GOOD TEACHER IS SEEN AS ONE WHO GIVES THE RIGHT ANSWERS TO THE STUDENTS IN A STRAIGHT FORWARD MANNER. THE TEST OF LEARNING IS TO GIVE BACK THE RIGHT ANSWERS. INFORMATION IS DEFINITIONS, CONCEPTS AND PARTS OF THINGS WITH LITTLE OR NO COMPARISON, ANALYSIS OR SYNTHESIS OF INFORMATION. THE DUALIST PREFERS A CLASSROOM ATMOSPHERE THAT IS PERSONAL, SUPPORTIVE AND NON-THREATENING. THE DUALIST LIKES A COURSE WITH A HIGH AMOUNT OF STRUCTURE, WHERE EXAMPLES ARE CONCRETE. AN ENVIRONMENT WHERE PEOPLE ARE LISTENED TO AND STUDENTS HAVE CHANCES TO PRACTICE NEW SKILLS. THE TEACHER IS SEEN AS A HELPER BUT NOT TOO PERSONAL.

IN ORDER TO MOVE STUDENTS FROM ONE PERRY STAGE TOWARD THE NEXT THEY NEED TO HAVE EXPERIENCES WHICH CAUSE SOME LEVEL OF DISCOMFORT, CONFLICT, UNCERTAINTY, AND AMBIGUITY. THE STUDENTS' MIND SET MUST BE CHALLENGED. THESE CHALLENGES SHOULD BE INTRODUCED A LITTLE AT A TIME WITH THE TEACHER DIFFUSING SOME OF THE RESULTING DISCOMFORT. THIS IS SLOW GOING AS TOO MUCH CHALLENGE CAN CAUSE STUDENTS TO RETREAT OR REVERT TO A PREVIOUS STAGE. THE DISCOMFORT CAN BE DIFFUSED BY USING A TEACHING STYLE THAT IS INFORMAL AND PERSONALLY RESPONSIVE TO THE STUDENTS, ALLOWS THE STUDENTS TO SHARE AND WHERE THE TEACHER SHARES THEIR OWN EXPERIENCES. I TRY TO MAKE THE CLASS AS NON-THREATENING AS POSSIBLE. STUDENTS AT DIFFERENT STAGES HAVE DIFFERENT PREFERRED LEARNING STYLES AND THEREFORE RESPOND DIFFERENTLY TO TEACHING STYLES. IF THERE IS A MATCH BETWEEN THE STUDENTS' LEARNING STYLES AND THE INSTRUCTOR'S TEACHING STYLE EVERYONE IS AT EASE, HOWEVER, A MISMATCH BETWEEN LEARNING AND TEACHING STYLE RESULTS IN ANGER, FRUSTRATION, ANXIETY AND HOSTILITY FOR ALL PARTIES. THIS IS ESPECIALLY TRUE FOR THE DUALISTIC STUDENTS. SINCE

DUALISTIC STUDENTS WANT PRECISE DIRECTION WITH LOTS OF STRUCTURE ANY INCREASE IN FREEDOM OR UNCERTAINTY IS VERY DIFFICULT FOR THEM TO HANDLE. BY USING THE PERRY SCHEMA I CAN BETTER MATCH MY TEACHING STYLE TO THE STUDENTS' LEARNING STYLES AND CAN PLAN ACTIVITIES TO HELP STUDENTS DEVELOP OR PROGRESS FROM THE DUALIST TOWARDS THE MULTIPLICITY OR RELATIVISTIC STAGES.

MOST OF MY GENERAL BIOLOGY I AND II CLASSES ARE COMPOSED OF FRESHMAN BIOLOGY MAJORS JUST OUT OF HIGH SCHOOL BUT SOMETIMES I TEACH THE NIGHT SECTION WHICH IS COMPOSED PRIMARILY OF OLDER STUDENTS WHO ARE GOING PART TIME AND/OR ARE JUST RETURNING TO COLLEGE. I WOULD EXPECT MANY OF THE OLDER STUDENTS TO BE FURTHER ALONG THE PERRY SCHEMA, BUT VERY OFTEN I HAVE FOUND THAT THESE STUDENTS IN TERMS OF BIOLOGY ARE DUALISTIC EVEN THOUGH IN TALKING WITH THEM THEY ARE MULTIPLICITISTS OR RELATIVISTS IN OTHER AREAS. THE FOLLOWING IS ONE EXAMPLE OF HOW PERRY AND OTHERS HAVE CHANGED THE WAY I TEACH. I HAVE ALWAYS BEEN AN HIGHLY ORGANIZED TEACHER BUT I DISCOVERED THAT I WASN'T GIVING THE DUALISTIC STUDENTS ENOUGH STRUCTURE. FOR EXAMPLE, I WOULD GIVE A VERBAL LABORATORY HOMEWORK ASSIGNMENT I.E. GRAPH THE RESULTS OF THE EXPERIMENT AND HAND IT IN NEXT WEEK WITH A BRIEF DISCUSSION. NOW I GIVE THE STUDENTS WRITTEN ASSIGNMENTS THAT ARE FULLY DEVELOPED (SPELLED OUT). CONSEQUENTLY THE WORK I RECEIVE IS MUCH IMPROVED. I WILL DISCUSS HOW TO PLAN A WRITING ASSIGNMENT LATER (SEE CHAPTER 5 ON DESIGNING WRITING ASSIGNMENTS).

WOMEN'S WAYS OF KNOWING

THE WORK OF BELENKY, CLINCHY, GOLDBERGER AND TARULE (1986) GIVES INSIGHT INTO AND AN UNDERSTANDING OF WOMEN'S EXPERIENCE, HOW THEY EXPERIENCE KNOWLEDGE AND HOW THEY DEVELOP. THE AUTHORS HAVE GROUPED "WOMEN'S WAYS OF KNOWING" INTO FIVE CATEGORIES;
    SILENCE: WHERE WOMEN SEE THEMSELVES AS VOICELESS AND
        POWERLESS
    RECEIVED KNOWLEDGE: WHERE WOMEN LISTEN AND RECEIVE KNOWLEDGE
        FROM OTHERS (AUTHORITIES) BUT DO NOT CREATE
        KNOWLEDGE
    SUBJECTIVE KNOWLEDGE: WHERE WOMEN HAVE AN INNER VOICE, THEY
        KNOW ON THE PERSONAL LEVEL (SUBJECTIVELY)
    PROCEDURAL KNOWLEDGE: WHERE WOMEN APPLY THE VOICE OF REASON,
        PROBLEM SOLVING OR PRACTICAL ANALYSIS TO OBTAINING
        KNOWLEDGE
    CONSTRUCTED KNOWLEDGE: WHERE WOMEN HAVE AN INTEGRATED VOICE
        AND ARE REFLECTIVE, KNOWLEDGE IS SEEN AS CONTEXTUAL
        AND BOTH OBJECTIVE AND SUBJECTIVE METHODS OF
        ACQUISITION ARE USED.
THIS WORK RAISES THE FOLLOWING QUESTIONS: HOW CAN WE HELP WOMEN DEVELOP? WHAT KIND OF SUPPORT AND/OR ENVIRONMENT IS NEEDED? WHAT IS NEEDED IN THE CLASS ROOM TO ACCOMPLISH GROWTH?

THERE ARE SIMILARITIES BETWEEN SOME OF THE WOMEN'S WAYS OF

KNOWING CATEGORIES AND PERRY STAGES OF DEVELOPMENT E.G. THE
DUALISTIC STAGE AND THE RECEIVED KNOWLEDGE CATEGORY, THE
MULTIPLICITY STAGE AND SUBJECTIVE KNOWLEDGE CATEGORY. HOWEVER
OVERALL THE WAYS THAT MEN DEVELOP AND THAT WOMEN KNOW ARE
DIFFERENT. JUST AS IN THE PERRY'S SCHEMA, IN EACH OF THE
DIFFERENT CATEGORIES OF WOMEN'S WAYS OF KNOWING, STUDENTS HAVE A
SPECIFIC RELATIONSHIP TO EDUCATION, KNOWLEDGE, AND FACTS OR
NEEDS.

FOR WOMEN TO DEVELOP THEY NEED TO KNOW THAT THEY HAVE
INTELLIGENCE, ARE COMPETENT, THAT PERSONAL EXPERIENCE IS A SOURCE
OF LEGITIMATE KNOWLEDGE AND HAS VALUE, THAT THERE IS A
COMMONALITY OF EXPERIENCE, AND THAT KNOWLEDGE IS CONNECTED.
THESE IDEAS NEED TO BE ACKNOWLEDGED AND ESTABLISHED BEFORE WOMEN
CAN REALLY LEARN AND DEVELOP. THE LEARNING ENVIRONMENT SHOULD BE
SUPPORTIVE, POSITIVE, NON-CONFRONTORY WITH A SENSE OF COMMUNITY.

BY USING PEER GROUPS WHICH DEMONSTRATE THE VALIDITY OF
DIFFERENT PERSPECTIVES AND EXPERIENTIAL/PERSONAL KNOWLEDGE AN
ENVIRONMENT OF ACCEPTANCE, SUPPORT AND COMMUNITY CAN BE
DEVELOPED. THE USE OF ANCHORING EXERCISES AND POSITIVE,
SUPPORTIVE COMMENTS BY PEERS AND FACULTY SIGNALS THE ACCEPTANCE
OF PERSONAL EXPERIENCE AS A SOURCE OF KNOWLEDGE. A TEACHER WHO
HAS A "COURTEOUS" ATTITUDE AND WHO CAN BE A ROLE MODEL FOR A
SCHOLAR AS A REAL AND IMPERFECT PERSON AND WHO CAN DEMONSTRATE
THAT LEARNING IS AN ATTAINABLE ACTIVITY IS IMPORTANT. SINCE
WOMEN VIEW KNOWLEDGE AS CONNECTED AN EMPHASIS ON CONNECTEDNESS
AND PATTERN FORMULATION SHOULD BE FOSTERED BY CLASS DISCUSSION
AND PEER GROUP INTERACTION.

Understanding and Language Usage

IN 1986 JOE WILLIAMS GAVE A SEMINAR AT WILLIAM PATERSON
COLLEGE ON HIS IDEAS OF THE NOVICE VERSUS THE PROFESSIONAL
(EXPERT). HIS THEORY IS THAT STUDENTS APPROACH A PROBLEM FROM
WHERE THEY STAND IN RELATIONSHIP TO IT (THEIR BACKGROUND) AND THE
SOCIAL SITUATION IN WHICH IT IS EMBEDDED. HE CLASSIFIES
STUDENTS IN RELATIONSHIP TO A FIELD AS BEING PRESOCIALIZED (THE
LANGUAGE, CONCEPTS, STYLES AND CONNECTIONS OF THE DISCIPLINE ARE
ALL NEW, THE STUDENTS HAVE NO VOCABULARY, NO HISTORY IN THE
SUBJECT MATTER AND ARE INSECURE) TO SOCIALIZED (UNDERSTAND THE
DISCIPLINE, SOUND LIKE INSIDERS IN LANGUAGE USAGE) TO POST-
SOCIALIZED (EXPERTS, HAVE THEIR OWN VOICE AND CONVENTIONS, SOUND
LIKE INSIDERS BUT DO NOT HAVE TO, RELATE THEIR FIELD TO OTHER
DISCIPLINES). A PERSON MAY BE SOCIALIZED IN ONE FIELD AND PRE-
SOCIALIZED IN ANOTHER FIELD. THE LEVEL OF SOCIALIZATION IS NOT
TRANSFERABLE.

THE COMBINATION OF WILLIAM'S IDEAS ON UNDERSTANDING AND
LANGUAGE USAGE WITH PERRY'S IDEAS ON COGNITIVE DEVELOPMENT AND
ITS IMPLICATIONS ON LEARNING STYLE AND TEACHING STYLE AND
BELENKY'S ET AL. VIEWS ON WOMEN'S WAYS OF KNOWING HAVE HELPED ME

TO DESIGN MY COURSE STRATEGY. DESPITE THE STUDENTS' OVERALL
PERRY POSITIONS IN TERMS OF BIOLOGY I AM DEALING PRIMARILY WITH
PRE-SOCIALIZED STUDENTS REGARDLESS OF AGE (FRESHMAN OR RETURNING
STUDENT). I NOW SEE WHY SOMETHINGS DO OR DON'T WORK IN CLASS AND
HOW TO IMPROVE MY TEACHING TO ACCOMPLISH WHAT I WANT IN TERMS OF
STUDENT LEARNING AND DEVELOPMENT. I CAN ADDRESS BOTH THE ISSUE
OF DEVELOPMENT AND OF THE FOCUS OF THE ASSIGNMENT. I CAN OFFER
MATERIAL TO FIT WHERE THE STUDENTS ARE AND AT THE SAME TIME OFFER
EXPERIENCES TO GIVE THEM OPPORTUNITY FOR GROWTH.

OTHER IDEAS ON LEARNING AND KNOWING

SEVERAL OTHER THEORIES HAVE ADDED TO MY TEACHING FRAMEWORK
AND I ACKNOWLEDGE THESE IDEAS EITHER DIRECTLY TO MY STUDENTS IN
LECTURE/DISCUSSION OR HAVE INCORPORATED THEM IN THE OVERALL
COURSE DESIGN. THE IDEA THAT STUDENTS HAVE A PERSONAL PREFERRED
LEARNING STYLE AND THAT THIS WILL TO SOME EXTENT MATCH OR
MISMATCH WITH THE DEMANDS OF DIFFERENT DISCIPLINES (KOLB, 1981)
IS AN IMPORTANT CONCEPT FOR ME AND THE STUDENTS TO UNDERSTAND.
KOLB SEPARATES LEARNING STYLES INTO TWO BASIC DIMENSIONS;
ABSTRACT TO CONCRETE AND ACTIVE TO REFLECTIVE. I TALK DIRECTLY
ABOUT THE IDEA THAT DIFFERENT STUDENTS LEARN DIFFERENTLY DURING
MY INTRODUCTORY LECTURE (SEE APPENDIX III, ASSIGNMENT # 31). I
DON'T GO INTO THE TERMINOLOGY BUT USE THE IDEA TO EXPLAIN WHY FOR
SOME STUDENTS SIENCE IS EASY TO LEARN WHILE FOR OTHERS IT IS
MORE DIFFICULT. I DISCUSS THIS TO HELP ALLEVIATE SOME OF THE
SCIENCE ANXIETY SUFFERED BY MANY STUDENTS. THE DISCUSSION SEEMS
TO FREE THEM UP. I ALSO EXPLAIN THAT EVEN THOUGH THEY HAVE A
PREFERRED LEARNING STYLE THEY ALSO HAVE CERTAIN STRATEGIES FOR
ACTUALLY WORKING WITH A PROBLEM AND THAT THESE STRATEGIES
OVERRIDE THE PREFERRED STYLE. ONE OF THE THINGS THAT WE DEAL
WITH IN GENERAL BIOLOGY IS THE SCIENTIFIC WAY OF APPROACHING
PROBLEMS OR WHAT STRATEGIES SCIENTISTS USE TO SOLVE PROBLEMS.

ONCE STUDENTS REALIZE THAT THEY HAVE DIFFERENT LEARNING
STYLES IT IS EASY TO MOVE ON TO THE IDEA THAT JUST AS THEY LEARN
DIFFERENTLY THERE ARE DIFFERENT PLACES TO PUT "KNOWING" THINGS
INTO THEIR BRAIN OR THAT THEY CAN KNOW THINGS IN DIFFERENT WAYS
(FORMAN AND GRASHA, 1983). I POINT OUT THAT THE MORE WAYS THEY
KNOW SOMETHING THE BETTER CHANCE THEY HAVE OF REMEMBERING IT.
SINCE IT TURNS OUT THAT MOST STUDENTS DON'T KNOW HOW THEY LEARN
BEST, I STRUCTURE MY COURSE SO THEY CAN EXPERIENCE ALL THE WAYS
OF KNOWING TO SEE WHAT WORKS BEST FOR THEM. THE WAYS OF KNOWING
ARE BY HEARING (LISTENING), READING (SEEING), WRITING, SPEAKING
AND DRAWING (IMAGING).

MY GENERAL BIOLOGY CLASS IS COMPOSED PRIMARILY OF PRE-
SOCIALIZED STUDENTS AND ALMOST EVERYTHING IS NEW BOTH IN TERMS OF
CONCEPTS AND VOCABULARY. BERNICE BRAID (1986) SAYS; TELL ME - I
FORGET, SHOW ME - I REMEMBER, AND INVOLVE ME - I UNDERSTAND.
STUDENTS LEARN BY DOING (BEING INVOLVED), BY HAVING AN ACTIVE
ROLE, BY CREATING OWNERSHIP OF THE INFORMATION OR EXPERIENCE. I

HAVE ALWAYS KNOWN THAT THIS WAS IMPORTANT IN LABORATORY BUT NOW I INVOLVE STUDENTS IN A MORE ACTIVE OR INTERACTIVE ROLE IN LECTURE. WHEN STUDENTS WORK WITH THE INFORMATION THEY ARE REALLY TRANSFORMING IT INTO THEIR OWN EXPERIENCE. THEY CAN CONSCIOUSLY RELATE IT TO THEIR OWN EXISTING KNOWLEDGE OR PRECONCEPTIONS. THEY CAN MAKE CONNECTIONS AND DISTINCTIONS WHEN THEY TRANSFORM INFORMATION INTO THEIR OWN WORDS.

LEARNING IS THE ACQUISITION OF NEW INFORMATION AND BY WORKING WITH THE MATERIAL THE STUDENTS TRANSFORM IT INTO A DIFFERENT WAY OF KNOWING. THE NORMAL WAY FOR STUDENTS TO TAKE IN INFORMATION IS BY LISTENING (LECTURE) AND BY READING (TEXT ASSIGNMENTS) AFTER WHICH THEY MUST TRANSFORM IT IN ORDER TO LEARN. THE TRANSFORMATION CAN BE BY WRITING, SPEAKING OR BY DRAWING. STUDENTS TAKE IN INFORMATION IN THE FOLLOWING WAYS:

1. HEARING- STUDENTS LISTEN TO LECTURES. HOWEVER, WHAT I SAY AND WHAT THEY HEAR CAN BE AND MOST OFTEN IS VERY DIFFERENT.
2. READING- STUDENTS READ THE PRINTED WORDS AND THEN MUST INTERPRET OR MAKE SENSE OUT OF IT. WHAT SENSE THEY MAKE WILL DEPEND ON THEIR OWN PARTICULAR FRAME OF REFERENCE.
3. WRITING- BY WRITING STUDENTS HAVE TO CONVERT WHAT THEY HAVE READ, SEEN OR HEARD INTO THEIR OWN WORDS. THEY MUST WORK WITH THE INFORMATION (SEE CHAPTER 5 FOR FURTHER DISCUSSION).
4. SPEAKING- STUDENTS INTERACT WITH THE INFORMATION AND TRANSFORM INTO THEIR OWN LANGUAGE WHEN THEY EXPLAIN OR TEACH THE INFORMATION TO ANOTHER STUDENT.
5. DRAWING (GRAPHIC REPRESENTATION OR ILLUSTRATION OR IMAGING)- STUDENTS TRANSFORM THE INFORMATION INTO A VISUAL FORM, A GRAPH OR ILLUSTRATION. IMAGING PUTS THE INFORMATION IN LONG TERM MEMORY AND GIVES STUDENTS A DIFFERENT PERSPECTIVE. SEMANTIC AND MENTAL IMAGING HELP STORE INFORMATION AND MAKE CONNECTIONS EXPLICIT (FORMAN AND GRASHA, 1983). THE RESULT OF IMAGING IS GREATER OWNERSHIP, AIDED LEARNING AND BETTER RECALL.

I TRY TO INCORPORATE AS MUCH TRANSFORMATION ACTIVITY AS POSSIBLE IN LECTURE AND LABORATORY AND I TELL THE STUDENTS WHY THEY ARE DOING WHAT THEY ARE DOING. I CAN NOT STRESS ENOUGH THE IMPORTANCE OF TELLING STUDENTS WHY THEY ARE DOING SOMETHING. FOR YEARS I ASSUMED THAT STUDENTS KNEW THE WHY AND WOULD JUST GET IT BY OSMOSIS. NOW I TELL THEM SO WE BOTH CAN UNDERSTAND EXACTLY WHAT IS GOING ON AND WHY.

LANGUAGE, WRITING AND THINKING

ROBERT PARKER (1987) MAINTAINS THAT THE KINDS OF LANGUAGE STUDENTS USE INFLUENCES THE KINDS OF LEARNING AND THINKING THEY

ARE DOING. DIFFERENT KINDS OF LANGUAGE USE PRODUCE DIFFERENT KINDS OF THINKING. WRITING IS NOT ONLY A MEANS OF THINKING BUT A MANIFESTATION OF THE THOUGHT PROCESS. THE RANGE OF WRITING IS FROM THE FORMAL TO THE INFORMAL AND FROM ANALYTIC TO IMAGINATIVE. BY USING DIFFERENT TYPES OF WRITING STUDENTS USE DIFFERENT KINDS OF LANGUAGE AND THEREFORE EXPERIENCES DIFFERENT KINDS OF LEARNING AND THINKING. THE USE OF THE FULL SPECTRUM OF WRITING STYLES (TYPES) SUPPORTS VERBAL THINKING USE AND DEVELOPMENT. I HAVE ALWAYS MAINTAINED THAT ANY FIRST YEAR INTRODUCTORY COURSE IS PROBABLY THE MOST DIFFICULT AS THE STUDENTS HAVE TO LEARN A TREMENDOUS AMOUNT OF BASIC VOCABULARY, THEORY AND CONCEPTS. I TELL THE STUDENTS THAT SCIENTIFIC VOCABULARY IS LIKE LEARNING A FOREIGN LANGUAGE AND THAT THE MORE THEY USE IT AND THE MORE WAYS THEY USE IT, THE MORE WAYS THEY WILL UNDERSTAND IT AND THE EASIER IT WILL BE TO LEARN AND INCORPORATE IT INTO THEIR KNOWLEDGE BASE. TCHUDI (1986) SAYS THAT THE QUALITY OF STUDENT WRITING REFLECTS THEIR ABILITY TO HANDLE AND UNDERSTAND CONTENT. WHEN I READ STUDENT WRITING I ASK MYSELF DOES THIS WRITING COMMUNICATE LEARNING IN SCIENCE. PARKER'S IDEAS ON THE RELATIONSHIP BETWEEN LANGUAGE, LEARNING AND THINKING ARE USEFUL TO ME WHEN I CONSIDER WHAT KINDS OF ASSIGNMENTS TO DESIGN.

CHAPTER 4

ACADEMIC ABILITIES

CURRICULAR REFORM HAS BEEN FOCUSED FOR THE LAST FEW YEARS ON WHAT SKILLS, COMPETENCIES, OR ABILITIES STUDENTS SHOULD HAVE WHEN THEY GRADUATE FROM COLLEGE OTHER THAN JUST EXPERTISE IN THE MAJOR AREA OF STUDY. THE QUESTION ASKED IS WHICH ACADEMIC COMPETENCIES, ABILITIES OR SKILLS SHOULD STUDENTS BE IMPROVING OR ACQUIRING DURING FOUR YEARS OF COLLEGE? I HAVE READ, STUDIED AND BEEN PRESENTED WITH NUMEROUS LISTS OF ABILITIES, COMPETENCIES AND SKILL AND HAVE NOTED A REMARKABLE AMOUNT OF OVERLAP. WHAT FOLLOWS IS A COMPOSITE LIST OF THESE ABILITIES WHICH I USED WHEN I PLANNED GENERAL BIOLOGY I AND II. NOT ALL THE ABILITIES ARE GIVEN EQUAL EMPHASIS IN THIS COURSE BUT ALL ABILITIES ARE DIRECTLY OR INDIRECTLY USED SOMETIME DURING THE YEAR. EACH ASSIGNMENT IS DESIGNED TO FOCUS ON THE ABILITY BEING DEVELOPED AS WELL AS CONTENT. WHEN I DESIGN AN ASSIGNMENT I INTENTIONALLY POINT OUT TO THE STUDENTS THE ABILITIES THEY ARE PRACTICING BECAUSE I WANT THEM TO BE AWARE THAT EDUCATION IS NOT JUST FACTS OR CONTENT BUT A CONGLOMERATION OF ABILITIES THAT CAN BE USED IN ALL DISCIPLINES.

LIST OF ACADEMIC ABILITIES:
COMMUNICATION - ORAL AND WRITTEN
ANALYTICAL - EVALUATION - CRITICAL THINKING
PROBLEM SOLVING
VALUE FORMATION
SOCIAL INTERACTION
SYNTHESIS - INTEGRATION - SUMMATION
AESTHETIC APPRECIATION
AWARENESS OF OTHER CULTURES - SEEING FROM ANOTHER
PERSPECTIVE
ORGANIZATION - REFERENCING
CREATIVITY
DECISION MAKING
INTERPRETATION

SOME ACADEMIC ABILITIES ARE WHAT A FRIEND OF MINE REFERS TO AS META-ABILITIES I.E. THEY REQUIRE THE USE OF SEVERAL ABILITIES USED TOGETHER. SUCH META-ABILITIES INCLUDE CRITICAL THINKING AND PROBLEM SOLVING. BECAUSE CRITICAL THINKING AND PROBLEM SOLVING ARE SUCH COMPLEX PROCESSES AND BECAUSE THEY ARE NOW BEING WIDELY DISCUSSED THE FOLLOWING TWO SECTIONS CONTAIN SOME COMMENTS ON THEM AS WELL AS SPECIFIC REFERENCES FOR FURTHER READING.

CRITICAL THINKING

CRITICAL THINKING IS A COMPLEX PROCESS AND THEREFORE TAKES TIME TO DEVELOP. CHET MEYERS' (1986) BOOK IS AN EXCELLENT REFERENCE FOR ANYONE INTERESTED IN HOW TO STRUCTURE A CLASS TO PROMOTE CRITICAL THINKING. CRITICAL THINKING IS THE ABILITY TO

ANALYZE, PRIORITIZE, GIVE STRUCTURE TO INFORMATION, AND TO EXPAND
BEYOND OR TO USE THE INFORMATION IN A DIFFERENT FRAMEWORK.
CRITICAL THINKING IS A META-ABILITY WHICH IS DEVELOPED BY
PRACTICING THE SIMPLER STEPS IN ORDER TO PROGRESS TO THE MORE
COMPLEX OPERATIONS. TO TEACH CRITICAL THINKING THE TEACHER MUST
FIND A BALANCE BETWEEN LECTURE (CONTENT) AND STUDENT INTERACTION
WITH THE INFORMATION (PROCESS). AS STUDENTS INTERACT WITH THE
MATERIAL SEVERAL THINGS HAPPEN I.E. THEY BECOME AWARE THAT OTHERS
HAVE DIFFERENT IDEAS FROM THEMSELVES (IT BROADENS THEIR
PERSPECTIVE), THEY MAKE SENSE OUT OF NEW MATERIAL, OWNERSHIP OF
THE INFORMATION IS CREATED, STUDENTS EXPERIENCE DISEQUILIBRIUM
AND HAVE PRACTICE IN USING ACADEMIC ABILITIES. PERRY'S SCHEMA
OFFERS INSIGHT INTO WHY STUDENTS DO NOT THINK CRITICALLY AND HOW
A CLASS CAN BE STRUCTURED TO ASSIST IN TEACHING CRITICAL
THINKING.

SINCE GENERAL BIOLOGY IS A FRESHMAN COURSE IT IS HERE THAT
THE GROUNDWORK IS LAID FOR THE TEACHING OF CRITICAL THINKING IN
OTHER UPPER LEVEL COURSES. STUDENTS MUST FIRST BE SOCIALIZED IN
TERMS OF BIOLOGICAL VOCABULARY, CONCEPTS, AND METHODOLOGIES
BEFORE THEY CAN HANDLE THE MORE COMPLEX OPERATIONS OF CRITICAL
THINKING IN SCIENCE AND THIS SOCIALIZATION IS ONE OF THE MAJOR
PURPOSES OF GENERAL BIOLOGY. BY STRUCTURING THE PROCESS PART OF
THE COURSE THE FOUNDATION FOR CRITICAL THINKING CAN BE LAID AND
THE BEGINNING STEPS OF CRITICAL THINKING CAN BE PRACTICED.

I TRY TO ESTABLISH A CLASSROOM ATMOSPHERE OF DIALOGUE,
INTERCHANGE, AND PROBLEM SOLVING BY USING SMALL, FREQUENT WRITING
ASSIGNMENTS, SMALL GROUP ACTIVITIES, AND FULL CLASS DISCUSSION AS
WE WORK WITH THE MATERIAL. THIS APPROACH STARTS STUDENTS ON THE
PATH OF CRITICAL THINKING AS THEY MUST DEAL NOT ONLY WITH THE
COURSE CONTENT BUT WITH OTHER PERSPECTIVES AND IDEAS. I
INTRODUCE THE PROCESS OF SUMMARIZATION WHICH REQUIRES BOTH THE
TRANSFORMATION OF CONCEPTS/ISSUES AND THE SKILL OF PRIORITIZING.
I ALSO USE ANCHORING EXERCISES TO RELATE PRIOR LEARNING TO WHAT
IS GOING ON IN CLASS (SEE CHAPTER 7). I USE WRITING ASSIGNMENTS
AS WELL AS ESSAY QUESTIONS TO GIVE THEM PRACTICE IN THINKING
CRITICALLY.

AN ESSAY QUESTION CAN BE STRUCTURED TO ADDRESS THREE
DIFFERENT LEVELS OF CRITICAL THINKING:

1. LISTING OR GIVING THE FACTS E.G. TRACE THE CIRCULATORY
   PATHWAY OF BLOOD THROUGH THE HEART (PRIMARILY A
   DESCRIPTIVE LIST)

2. DISCUSSING THE SIGNIFICANCE/MEANING OF THE INFORMATION
   E.G. COMPARE THE CIRCULATORY PATHWAYS OF THE BLOOD
   THROUGH AN ADULT HEART AND A FETAL HEART. DISCUSS
   THE SIGNIFICANCE OF THE DIFFERENCES IN TERMS OF
   FETAL DEVELOPMENT. NOTE: WHAT THE WORDS COMPARE
   AND DISCUSS MEAN SHOULD BE MADE CLEAR TO THE STUDENTS

SO THAT THEY KNOW THAT WHAT YOU EXPECT IS NOT JUST A LIST OF ATTRIBUTES.

3. EVALUATING/GIVING AN OPINION OR FORMULATING POSSIBLE FUTURE CONSEQUENCES.

IN GENERAL BIOLOGY LECTURE I TRY TO ADDRESS THE SECOND LEVEL OF CRITICAL THINKING IN ALL ESSAY QUESTIONS. I STRUCTURE THE ESSAY QUESTIONS SO THAT THE STUDENTS ARE FORCED TO SYNTHESIZE INFORMATION. THE THIRD LEVEL CRITICAL THINKING DEVELOPMENT IS DEALT WITH PRIMARY IN LABORATORY WHEN STUDENTS CARRY OUT, DESIGN AND EVALUATE EXPERIMENTS, AND WRITE LABORATORY REPORTS. HOWEVER, WITHOUT PROPER PREPARATION, EXPLANATION AND DISCUSSION LABORATORY REPORTS ARE NOT USUALLY EXAMPLES OF A STUDENT'S CRITICAL THINKING ABILITY BUT ARE SIMPLY A REPHRASING OF THE LABORATORY MANUAL. I LEAD THE STUDENTS INTO CRITICAL THINKING BY USING A SEQUENCE OF SHORT ASSIGNMENTS WHICH WORK ON DIFFERENT ASPECTS OF CRITICAL THINKING AND THEN FINALLY HAVE THEM WORK THROUGH THE WHOLE PROCESS (SEE APPENDIX II).

PROBLEM SOLVING

PROBLEM SOLVING IS ANOTHER META-ABILITY AND IS ONE STRATEGY OR APPROACH USED TO HELP TEACH CRITICAL THINKING. FOR THOSE INTERESTED IN PROBLEM SOLVING AND THE USE OF WRITING AS A PROBLEM SOLVING PROCESS I STRONGLY RECOMMEND THE WORKS OF LINDA FLOWERS (1982) AND JOHN R.HAYES (1981). THE FOLLOWING PASSAGE IS FROM CHAPTER FIVE IN KLOSS (HANKS, 1985, PAGE 23). "PROBLEM SOLVING IS FIGURING OUT HOW TO GET FROM WHERE YOU ARE TO WHERE YOU WANT TO BE -- GETTING FROM A TO B. PROBLEM SOLVING COORDINATES MANY SEPARATE ABILITIES: IDENTIFYING RELATIONSHIPS AND MAKING OBSERVATIONS (ANALYSIS); APPLYING VALUES AT EACH STAGE OF DECISION MAKING (VALUEING); AND FORMULATING GOALS; AND ARTICULATING (RECOGNIZING) OBSTACLES AND IDENTIFYING COSTS AND ALTERNATE APPROACHES. PROBLEM SOLVING DEPENDS ON DEVELOPED SKILLS IN COMMUNICATION AND SOCIAL COMMUNICATION."

THERE ARE SPECIFIC STAGES TO PROBLEM SOLVING AND A NUMBER OF STEPS THAT CAN BE USED. STUDENTS SHOULD BE MADE AWARE OF THE STAGES AND STEPS NECESSARY TO SOLVE A PROBLEM AND KNOW THE USES AND ADVANTAGES OF EACH STAGE AND STEP.

IN GENERAL BIOLOGY I USE PROBLEM SOLVING MOST EXTENSIVELY IN LABORATORY, E.G.:

1. LABORATORY EVALUATIONS CAN BE USED TO GIVE STUDENTS EXPERIENCE IN PIN-POINTING THE PROBLEM. WHY (FOR WHAT PURPOSE) WAS THE EXPERIMENT DONE? THE ANSWER IS LIMITED TO TWO OR THREE SENTENCES (SEE APPENDIX I, ASSIGNMENT # 3).

2. LABORATORY REPORTS ARE MORE EXTENSIVE PROBLEM SOLVING

ASSIGNMENTS IN THAT THE STUDENTS ARE ASKED NOT ONLY TO
FIND THE PROBLEM BUT TO ORGANIZE THE IDEAS, IDENTIFY
ASSUMPTIONS, FORMULATE HYPOTHESIS, ORGANIZE DATA,
AND DRAW CONCLUSIONS (SEE APPENDIX II, ASSIGNMENT #
11).

CHAPTER 5

WRITING

WHY WRITE?

IN MY CLASSES I USE WRITING WITH AND WITHOUT PROCESS (SMALL GROUP INTERACTION AND CLASS DISCUSSION) TO WORK ON STUDENT COGNITIVE DEVELOPMENT, IMPROVE ACADEMIC ABILITIES, AND TO INCREASE CONTENT LEARNING. THERE ARE THREE KINDS OF WRITING:

1. TRANSACTIONAL- WRITING FOR SOMEONE ELSE, TO COMMUNICATE, USED AS AN EVALUATIVE TOOL. ALSO CALLED EXPOSITORY OR ANALYTICAL WRITING.
2. POETIC- ARTISTIC WRITING.
3. EXPRESSIVE- WRITING AS THINKING ON PAPER, REVEALS THE THINKING PROCESS, ALLOWS WRITERS TO FIND OUT WHAT THEY KNOW, WRITING TO AND FOR ONESELF, A PERSONAL EXPERIENCE, WRITING TO LEARN, THE FIRST STAGE TOWARDS TRANSACTIONAL WRITING, YOU CAN THINK THROUGH A PROBLEM.

UNTIL I BECAME INVOLVED IN THE WRITING ACROSS THE CURRICULUM PROGRAM I PRIMARILY ASKED FOR AND RECEIVED TRANSACTIONAL WRITING - IN THE FORM OF ANSWERS TO ESSAY EXAM QUESTIONS AND FINALIZED LABORATORY REPORTS. NOW THE MAJORITY OF WRITING I ASSIGN IS EXPRESSIVE (WRITING TO LEARN) WITH SOME TRANSACTIONAL WRITING (WRITING TO COMMUNICATE). I DO NOT IDENTIFY FOR STUDENTS WHAT KIND OF WRITING THEY ARE USING AS THE KINDS OF WRITING SOMETIMES SEEM TO RUN TOGETHER. THE IMPORTANT THING IS THAT I NOW EXTENSIVELY USE WRITING AS PROCESS IN MY CLASSES AND THE QUESTION IS WHY? WHY USE WRITING IN A CONTENT COURSE?

WRITING IS A THINKING PROCESS (COGNITIVE) (FLOWER, 1982), A SERIES OF CONCEPTUAL DECISIONS. IT ENGAGES BOTH VISUAL AND MOTOR PROCESSES, REINFORCES MEMORY, INCREASES SYNTHESIS OF INFORMATION AND READING SKILLS. WRITING FORCES STUDENTS TO BE INTERACTIVE LEARNERS AND IS CENTRAL TO SELF-DISCOVERY (JUDY AND JUDY, 1981). WRITING ALLOWS US TO KNOW WHAT WE KNOW. AS THE STUDENTS WRITE THEY ARE TRANSFORMING NEW INFORMATION, MANIPULATING IT TO FIT THE TASK, MAKING NEW CONNECTIONS, NEW COMBINATIONS, AND ARE EVALUATING AND ORDERING THE INFORMATION. WRITING MAKES THE STUDENTS REFLECT ON AND ANALYZE INFORMATION. BY WRITING STUDENTS GET TO DISCOVER FOR THEMSELVES AND GAIN NEW INSIGHT. STUDENTS ARE MAKING THEIR OWN MEANING BY INTERACTING WITH THE INFORMATION (WHITE, 1986). WHEN STUDENTS WRITE THEY ARE DISCOVERING WHAT THEY THINK. THEY KNOW WHAT THEY KNOW AND CREATE OWNERSHIP OF THE INFORMATION. WRITING IS A TOOL OF LEARNING. IT IS A VISIBLE ACCOUNT TO WHICH THE STUDENTS CAN REFER.

WRITING PROMOTES OR CAN BE USED TO TEACH ACTIVE LEARNING, CITICAL THINKING, PROBLEM SOLVING, AND ALL THE OTHER ACADEMIC ABILITIES. WRITING FOSTERS CLASSROOM DISCUSSION AND CAN BE USED

TO DEVELOP RAPPORT BETWEEN STUDENTS, AND BETWEEN THE STUDENTS AND THE TEACHER. IT INCREASES THE STUDENTS' EASE WITH NEW MATERIAL, REDUCES ANXIETY AND BROADENS THEIR POINT OF VIEW. WRITING CAN BE USED TO FOCUS STUDENT THINKING, INCREASE STUDENT CREATIVITY AND STIMULATE THE USE OF THEIR IMAGINATIONS.

WRITING IS INTENSELY FUNCTIONAL IN THAT DURING THE PROCESS STUDENTS ARE CHALLENGED TO REVIEW AND REFLECT ON IDEAS (BE REFLECTIVE) AND TO ANALYZE CONCEPTS AND SEE RELATIONSHIPS (BE ANALYTICAL). STUDENTS MUST MAKE USE OF THEIR KNOWLEDGE, REORGANIZE IT AND RETHINK IT TO MEET THE DEMANDS OF THE ASSIGNMENT. A WELL PLANNED WRITING ASSIGNMENT DEEPENS THE STUDENTS' UNDERSTANDING OF THE FIELD, AT THE SAME TIME IT HELPS STUDENTS CREATE MEANING OUT OF RAW DATA. WRITING GIVES CLARITY TO THE THOUGHT PROCESS AS THE WRITER EXPRESSES AND EXPLAINS MEANING. MUCH OF WHAT WE WRITE IS AN ATTEMPT TO PINPOINT OR DEFINE A PROBLEM, TO ZERO IN ON THE KEY ISSUES INVOLVED AND TO HELP FIND A SOLUTION.

WRITING CAN HELP STUDENTS ACTIVELY EXPERIENCE A NEW WAY OF DOING THINGS. THE ACTUAL 'DOING' IS IMPORTANT SO THAT THE STUDENTS WILL 'KNOW' (BE ABLE TO DO) RATHER THAN JUST 'KNOW' (RATIONALLY UNDERSTAND). WRITING MAY ALLOW OR BE STRUCTURED TO ALLOW DETACHING I.E. LOOKING AT AN IDEA OR PROBLEM FROM DIFFERENT POINTS OF VIEW OR PERSPECTIVES. DETACHING MAKES IT EASIER FOR STUDENTS TO RELAX ABOUT THE TOPIC AS THEY ARE NOT SO PERSONALLY INVOLVED. THEY CAN THEN REPHRASE THE WORK OR TOY WITH DIFFERENT IDEAS AS THEY ARE NOT DEFENDING THEIR POINT OF VIEW. DETACHING CAN BE A PRELUDE TO CREATIVE INSIGHTS OR INSPIRATIONS. WRITING IS A PROCEDURE FOR UNCOVERING NEW IDEAS AND DISCOVERING NEW CONFIGURATIONS OF OLD IDEAS. STUDENTS CAN EXPLORE THESE NEW CONFIGURATIONS AND CONNECTIONS MORE FREELY WHEN THEY ARE DETACHED.

TYPES OF WRITING ASSIGNMENTS

WHEN THINKING ABOUT WHAT KIND OF WRITING ASSIGNMENTS YOU WANT TO USE FOR A SPECIFIC CLASS, EXPERIMENT WITH DIFFERENT TYPES. NOT ALL TYPES OF ASSIGNMENTS ARE SUITABLE FOR ALL CLASSES BUT USUALLY SOME VARIETY IS POSSIBLE AND ADVISABLE. I USE VARIOUS TYPES OF ASSIGNMENTS TO FIT MY COURSE OBJECTIVES (SEE CHAPTER 12).

EXAMPLES OF DIFFERENT TYPES OF WRITING ASSIGNMENTS:
TERM OR RESEARCH PAPERS (LONG OR SHORT)
LONG REPORTS
*ESSAYS
MICROTHEMES
CRITICAL SUMMARIES OF ASSIGNED READING
*LETTERS
*QUESTIONS AND ANSWERS FOR EXAMINATIONS
*JOURNALS

```
*LOGS
*SHORT REPORTS
*LABORATORY REPORTS OR EVALUATIONS
 CRITIQUES OF PLAYS OR ART WORKS
 QUIZZES
 CASE STUDIES
*FREE WRITING
*CLASS PROJECTS (INDIVIDUAL, GROUP OR CLASS)
```

THE ASTRISKS (*) INDICATE THE TYPES OF WRITING ASSIGNMENTS THAT I AM PRESENTLY USING IN GENERAL BIOLOGY. EACH TYPE OF WRITING ASSIGNMENT IS BEST SUITED TO ACCOMPLISH DIFFERENT COURSE OBJECTIVES. WITH A VARIETY OF ASSIGNMENTS THE STUDENTS AND I DO NOT GET BORED WITH THE FREQUENCY OF WRITING.

DESIGNING WRITING ASSIGNMENTS

WHEN I FIRST STARTED USING WRITING IN MY CLASS I GAVE ORAL NOT WRITTEN INSTRUCTIONS. ALTHOUGH I HAD GIVEN THE ASSIGNMENT PRIOR CONSIDERATION, I HAD NOT REALLY PLANNED/DESIGNED IT. THE RESULTS CONSEQUENTLY RANGED FROM GOOD TO TERRIBLE. I THEN ATTENDED A WRITING ACROSS THE CURRICULUM WORKSHOP ON HOW TO DESIGN A WRITING ASSIGNMENT AND AS USUAL WITH MORE INSIGHT INTO THE PROCESS ON MY PART THE STUDENTS' RESPONSES STARTED TO IMPROVE.

ASSIGNMENTS REQUIRE PLANNING SO THAT THE ASSIGNMENT MESHES WITH THE OBJECTIVES OF THE COURSE (TCHUDI, 1986). THE ASSIGNMENTS MUST HAVE A CLEAR PURPOSE IF THEY ARE TO SERVE THE CURRICULAR DESIGN (WRITING ASSIGNMENTS MUST BE TAILORED FOR EACH COURSE). HERE ARE SOME AREAS TO BE CONSIDERED WHEN PLANNING AN ASSIGNMENT:

1. PURPOSE - WHAT IS THE PURPOSE OF THE ASSIGNMENT FOR THE STUDENTS AND FOR YOU? THERE MAY BE MORE THAN ONE PURPOSE.
2. HOW DOES THE ASSIGNMENT FIT IN WITH THE OBJECTIVES OF THE COURSE AND HOW DOES IT RELATE TO OTHER ASSIGNMENTS?
3. WHAT IS THE AUDIENCE? - MOST OFTEN WRITING IS WRITTEN TO YOU (THE AUTHORITY, THE TEACHER) BUT STUDENTS FREQUENTLY WRITE MORE FREELY IF THE AUDIENCE IS CHANGED FROM AN ALL-KNOWING FIGURE TO ANOTHER LESS KNOWLEDGEABLE AUDIENCE I.E. A PEER, RELATIVE OR THE SELF. WHEN STUDENTS WRITE TO A NON-AUTHORITY THEY CAN NOT ASSUME ANYTHING.
4. WHAT IS THE ASSIGNMENT? MAKE IT CLEAR AND TO THE POINT.
5. HOW DO I PLAN TO IMPLEMENT THE ASSIGNMENT IN CLASS?
6. WHAT BASIC INFORMATION IS NEEDED? MAKE IT CLEAR E.G. DUE DATES, KIND OF ASSESSMENT/EVALUATION THAT WILL BE USED (IF ANY), ASSESSMENT CRITERIA, SCORING GUIDE, NUMBER OF WORDS OR PAGES REQUIRED, TYPED OR WRITTEN ETC. (NOTE: NOT ALL ASSIGNMENTS NEED TO BE EVALUATED BY

YOU) (SEE CHAPTER 9).
7. DO THE ASSIGNMENT YOURSELF BEFORE YOU GIVE IT TO THE CLASS OR HAVE A SOMEONE ELSE TAKE IT (PRE-TEST).

EVEN AFTER I DO ALL MY PLANNING THERE ARE OR CAN BE SOME UNCLEAR AREAS IN AN ASSIGNMENT. TO TRY AND GET ALL THE KINKS OUT OF AN ASSIGNMENT PRIOR TO CLASS USE I HAVE FOUND A COUPLE OF COLLEAGUES WHO ARE WILLING TO PRE-READ OR PRE-TEST MY ASSIGNMENTS AND GIVE SUGGESTIONS AND COMMENTS. I DO THE SAME FOR THEM. THIS INTERACTION IS VERY REWARDING BOTH PROFESSIONALLY AND PERSONALLY.

8. GO OVER THE ASSIGNMENT WITH THE CLASS USING A HANDOUT. BE SURE TO ALLOW ENOUGH TIME FOR DISCUSSION SO THAT THE STUDENTS UNDERSTAND ALL ASPECTS OF THE ASSIGNMENT. WHEN THE ASSIGNMENT IS GIVEN TO THE STUDENTS IT WILL BE MORE SUCCESSFUL IF IT CLEARLY SPECIFIES THE WRITING TASK, THE AUDIENCE TO BE ADDRESSED AND THE ASSESSMENT/EVALUATION CRITERIA TO BE USED. A HANDOUT NOT ONLY HELPS THE STUDENTS TO FOCUS THEIR WORK AND INSURES THAT AT LEAST ALL THE STUDENTS HAVE THE SAME INFORMATION BUT HELPS US CLARIFY FOR OURSELVES THE STRUCTURE AND PURPOSE OF THE ASSIGNMENT. OFTEN WE RECEIVE WORK THAT IS OFF THE ASSIGNMENT BECAUSE WE WERE NOT CLEAR WHEN WE TALKED ABOUT THE ASSIGNMENT IN CLASS. WHAT WE THOUGHT WE SAID AND WHAT THE STUDENTS THOUGHT THEY HEARD MAY VARY MARKEDLY. A HANDOUT GIVES BOTH YOU AND THE STUDENTS SOMETHING TO REFER TO AND TO CHECK AGAINST. IT IS ALSO USEFUL TO GIVE OUT GOOD AND BAD EXAMPLES OF THE ASSIGNMENT FROM PREVIOUS CLASSES FOR DISCUSSION.

9. IF STUDENTS ARE DOING THE ASSIGNMENT IN CLASS DO IT WITH THEM. EVEN THOUGH THE ASSIGNMENT HAS BEEN PRE-TESTED I DO IT IN CLASS WITH THE STUDENTS. BY DOING THE ASSIGNMENT IN THE CLASSROOM SITUATION I SOMETIMES GAIN NEW INSIGHTS AS TO HOW IT CAN BE IMPROVED OR WHERE THERE MAYBE PROBLEMS. YOU CAN SHARE YOUR EFFORTS WITH THE CLASS AS THEY SHARE THEIRS WITH EACH OTHER AND YOU.

ALTHOUGH I FIND THE ABOVE APPROACH TO DESIGNING AN ASSIGNMENT TIME CONSUMING, THE IMPROVEMENT IN MY CURRICULAR DESIGN AND THE ABILITY TO ACTUALLY ACCOMPLISHING WHAT I WANT IN CLASS IS WELL WORTH THE EFFORT AND TIME. I HAVE A SAYING THAT "THE SYSTEM IS EVERYTHING". WHAT I MEAN IS THAT CONTENT AND PROCESS ARE ALL TIED TOGETHER BY THE WRITING ASSIGNMENTS. THIS APPROACH TO TEACHING REQUIRES A LOT MORE PREPARATION TIME THAN TEACHING ONLY CONTENT. HOWEVER, I FEEL LIKE I'M NOW TEACHING FOR THE FIRST TIME.

WHEN I DESIGN AN ASSIGNMENT I USE A PLANNING WORKSHEET SO THAT I DON'T FORGET ANY OF THE PLANNING AREAS. I CONSIDER EACH PART OF THE PLANNING WORKSHEET EVEN IF I DECIDE NOT TO USE IT FOR A PARTICULAR ASSIGNMENT. THE FOLLOWING ARE THE ASSIGNMENT

PLANNING WORKSHEET, AN EXAMPLE OF THE WORKSHEET WITH AN
ASSIGNMENT, AND THE SAME ASSIGNMENT AS THE STUDENTS RECEIVE IT.

ASSIGNMENT PLANNING WORKSHEET

ASSIGNMENT PLANNING WORKSHEET

ASSIGNMENT NUMBER _____ TOPIC/LAB _____

1. PURPOSE:
   TEACHER:
   STUDENT:

2. HOW DOES THE ASSIGNMENT FIT WITH THE OBJECTIVES OF THE COURSE?

3. HOW DOES THE ASSIGNMENT RELATE TO OTHER ASSIGNMENTS?

4. WHAT IS THE AUDIENCE FOR THE ASSIGNMENT?

5. THE ASSIGNMENT.

6. HOW WILL I IMPLEMENT THE ASSIGNMENT IN CLASS?

7. DATE DUE _____ LENGTH (PAGES AND WORDS) _____
   TYPED ___ OR WRITTEN ___

8. WILL THERE BE ANY SUPPORTING MATERIALS OR ROUGH DRAFTS HANDED
   IN WITH THE ASSIGNMENT?_____

9. ASSESSMENT/EVALUATION

   HOW AM I GOING TO ASSESS/EVALUATE THIS ASSIGNMENT?
   CRITERIA:
   SCORING GUIDE:

10. HAVE I PRE-TESTED THE ASSIGNMENT? _____ HOW?_____

11. IS THE HANDOUT READY? _____

    DO I HAVE COPIES OF PREVIOUS EXAMPLES OF STUDENT RESPONSES TO
    THIS ASSIGNMENT TO HAND OUT IN CLASS? _____

AFTER THE ASSIGNMENT HAS BEEN DONE.

12. ARE THERE ANY CHANGES TO BE MADE IN THIS ASSIGNMENT?

13. DO I HAVE NEW EXAMPLES OF STUDENT RESPONSES FOR THIS
    ASSIGNMENT?
    YES_____ NO _____

ASSIGNMENT PLANNING WORKSHEET WITH AN ASSIGNMENT.

ASSIGNMENT PLANNING WORKSHEET

NUMBER _1__   TOPIC/LAB   LECTURE : LOG

1. PURPOSE:
   TEACHER: THIS ASSIGNMENT IS TO HELP ME TO GET TO KNOW THE
   STUDENTS, TO HELP THEM WITH PROBLEMS, TO SPOT
   STUDENTS WHO NEED TUTORING, AND TO KEEP ME IN TOUCH
   WITH ANY PROBLEMS THEY ARE HAVING WITH THE MATERIAL.
   STUDENT: THIS ASSIGNMENT IS TO HELP ANSWER YOUR QUESTIONS,
   RELIEVE YOUR ANXIETY ABOUT THE MATERIAL, EXPRESS YOUR
   FEELINGS, GET HELP, ALLOW REFLECTION ABOUT THE
   COURSE, YOURSELF AND LIFE IN GENERAL, AND TO VALIDATE
   YOUR PERSONAL EXPERIENCES.

2. HOW DOES THE ASSIGNMENT FIT WITH THE OBJECTIVES OF THE COURSE?
   THIS ASSIGNMENT HELPS TO ESTABLISH A SUPPORTIVE STRUCTURE
   FOR THE CLASS AND HELPS THE STUDENTS BECOME MORE ACTIVE
   LEARNERS.

3. HOW DOES THE ASSIGNMENT RELATE TO OTHER ASSIGNMENTS?
   NOT APPLICABLE

4. WHAT IS THE AUDIENCE FOR THE ASSIGNMENT?
   THE TEACHER

5. THE ASSIGNMENT.
                          LOG

   A LOG IS A 5 MINUTE WRITING EXERCISE DUE AT THE BEGINNING OF
THE SECOND LECTURE OF EACH WEEK. LOGS ARE VOLUNTARY.

   WHAT IS THE PURPOSE OF THE LOG EXPERIENCE? THE PURPOSE OF
THE LOG FOR ME IS SO THAT I CAN GET TO KNOW YOU BETTER AND
PERHAPS ANSWER ANY QUESTIONS YOU MAY HAVE DURING THE SEMESTER.
FOR THE CLASS THE LOGS CAN FOSTER A FREER EXCHANGE OF IDEAS AS
YOU (THE STUDENTS) MAY FEEL THAT IT IS EASIER TO ASK QUESTIONS
AND DISCUSS IDEAS. FOR YOU, PERSONALLY, THIS IS A CHANCE TO
REFLECT ON AND EXPRESS WHAT IS GOING ON IN YOUR LIFE.

   FIVE MINUTES OF WRITING IS APPROXIMATELY 1/2 TO 1 PAGE OF
WRITING (@ 150-250 WORDS). LOGS MAYBE TYPED OR WRITTEN BY HAND.
YOUR LOGS ARE NOT GRADED BUT I KEEP A RECORD OF HOW MANY YOU HAVE
HANDED IN DURING THE SEMESTER. LOGS ARE ENTIRELY CONFIDENTIAL
BETWEEN YOU AND ME. I RETURN YOUR LOG TO YOU AT THE BEGINNING OF
THE NEXT LECTURE SESSION.

   YOU MAY WRITE ON ANY SUBJECT. THE FOLLOWING ARE A FEW
EXAMPLES OF WHAT STUDENTS HAVE WRITTEN ABOUT IN PAST CLASSES:
QUESTIONS NOT ASKED IN CLASS, QUESTIONS ABOUT THE COURSE,

PRACTICE ESSAY ANSWERS FOR EXAMS, LECTURE SUMMARIES, QUESTIONS
ABOUT SCHOOL IN GENERAL, THOUGHTS ABOUT SCHOOL, HOME, JOBS,
SPORTS, HOBBIES, FOOD, POEMS AND SONGS WRITTEN BY STUDENTS,
THINGS YOU THINK ABOUT IN GENERAL.

IN OTHER WORDS LOGS CAN BE ANYTHING AT ALL.

HOW DO I USE THE LOGS IN TERMS OF COURSE EVALUATION? I USE
THE LOGS AS A SWING GRADE THAT IS IF YOU ARE ON THE BORDER
BETWEEN A C+ (79%) AND AN B- (80%) AND YOU HAVE HANDED IN 2/3 OF
THE LOGS DURING THE SEMESTER (E.G. 10 OUT OF 15) YOU WILL RECEIVE
THE B- BUT IF YOU HAVEN'T TURNED IN THE LOGS YOU GET THE C+.

IF YOU HAVE ANY QUESTIONS ABOUT THE LOGS PLEASE ASK ME.

6. HOW WILL I IMPLEMENT THE ASSIGNMENT IN CLASS?
   I GIVE OUT THE LOG DURING THE FIRST LECTURE SESSION AND
   DISCUSS IT WITH THE CLASS. I ALSO ASK IF THEY HAVE ANY
   QUESTIONS AT THE SECOND LECTURE WHEN THE FIRST LOGS CAN BE
   HANDED IN.

7. DATE DUE: SECOND LECTURE SESSION EACH WEEK
   LENGTH (PAGES AND WORDS): 1/2 TO 1 PAGE (150-250 WORDS)
   TYPED OR WRITTEN

8. WILL THERE BE ANY SUPPORTING MATERIALS OR ROUGH DRAFTS HANDED
   IN WITH THE ASSIGNMENT? NO

9. ASSESSMENT/EVALUATION

   HOW AM I GOING TO ASSESS/EVALUATE THIS ASSIGNMENT?
   LOGS ARE CHECKED OFF AND RECORDED AS THEY ARE HANDED IN.
   I MAKE SUPPORTIVE COMMENTS OR ANSWER QUESTIONS DEPENDING
   ON THE CONTENT OF THE LOG.

   CRITERIA: LENGTH

   SCORING GUIDE: A LOG RECEIVES A CHECK IF IT IS LONG ENOUGH.

10. HAVE I PRE-TESTED THE ASSIGNMENT? YES
    HOW? USED IT LAST YEAR

11. IS THE HANDOUT READY? YES

    DO I HAVE COPIES OF PREVIOUS EXAMPLES OF STUDENT RESPONSES TO
    THIS ASSIGNMENT TO HAND OUT IN CLASS? NO

AFTER THE ASSIGNMENT HAS BEEN DONE.

12. ARE THERE ANY CHANGES TO BE MADE IN THIS ASSIGNMENT?

13. DO I HAVE NEW EXAMPLES OF STUDENT RESPONSES FOR THIS ASSIGNMENT?

YES_____   NO _____

THE ASSIGNMENT AS THE STUDENT RECEIVES IT.

LOG

A LOG IS A 5 MINUTE WRITING EXERCISE DUE AT THE BEGINNING OF THE SECOND LECTURE OF EACH WEEK. LOGS ARE VOLUNTARY.

WHAT IS THE PURPOSE OF THE LOG EXPERIENCE? THE PURPOSE OF THE LOG FOR ME IS SO THAT I CAN GET TO KNOW YOU BETTER AND PERHAPS ANSWER ANY QUESTIONS YOU MAY HAVE DURING THE SEMESTER. FOR THE CLASS THE LOGS CAN FOSTER A FREER EXCHANGE OF IDEAS AS YOU (THE STUDENTS) MAY FEEL THAT IT IS EASIER TO ASK QUESTIONS AND DISCUSS IDEAS. FOR YOU, PERSONALLY, THIS IS A CHANCE TO REFLECT ON AND EXPRESS WHAT IS GOING ON IN YOUR LIFE.

FIVE MINUTES OF WRITING IS APPROXIMATELY 1/2 TO 1 PAGE OF WRITING (@ 150-250 WORDS). LOGS MAYBE TYPED OR WRITTEN BY HAND. YOUR LOGS ARE NOT GRADED BUT I KEEP A RECORD OF HOW MANY YOU HAVE HANDED IN DURING THE SEMESTER. LOGS ARE ENTIRELY CONFIDENTIAL BETWEEN YOU AND ME. I RETURN YOUR LOG TO YOU AT THE BEGINNING OF THE NEXT LECTURE SESSION.

YOU MAY WRITE ON ANY SUBJECT. THE FOLLOWING ARE A FEW EXAMPLES OF WHAT STUDENTS HAVE WRITTEN ABOUT IN PAST CLASSES: QUESTIONS NOT ASKED IN CLASS, QUESTIONS ABOUT THE COURSE, PRACTICE, ESSAY ANSWERS FOR EXAMS, LECTURE SUMMARIES, QUESTIONS ABOUT SCHOOL IN GENERAL, THOUGHTS ABOUT SCHOOL, HOME, JOBS, SPORTS, HOBBIES, FOOD, POEMS AND SONGS WRITTEN BY STUDENTS, THINGS YOU THINK ABOUT IN GENERAL.

IN OTHER WORDS LOGS CAN BE ANYTHING AT ALL.

HOW DO I USE THE LOGS IN TERMS OF COURSE EVALUATION? I USE THE LOGS AS A SWING GRADE THAT IS IF YOU ARE ON THE BORDER BETWEEN A C+ (79%) AND AN B- (80%) AND YOU HAVE HANDED IN 2/3 OF THE LOGS DURING THE SEMESTER (E.G. 10 OUT OF 15) YOU WILL RECEIVE THE B- BUT IF YOU HAVEN'T TURNED IN THE LOGS YOU GET THE C+.

IF YOU HAVE ANY QUESTIONS ABOUT THE LOGS PLEASE ASK ME.

CHAPTER 6

STRATEGIES

I USE SEVERAL DIFFERENT STRATEGIES OR TECHNIQUES IN CLASS TO GET THE STUDENTS TO INTERACT WITH THE COURSE CONTENT, THEMSELVES AND WITH ME. STUDENTS WORK INDIVIDUALLY, IN PEER GROUPS OF FOUR OR FIVE, WITH LABORATORY PARTNERS AND WITH THE CLASS AS A WHOLE. I USE DIFFERENT WRITING STRATEGIES INDIVIDUALLY AND IN COMBINATION.

BRAINSTORMING

THE PURPOSE OF BRAINSTORMING IS TO GENERATE IDEAS - - THE MORE THE BETTER. THE IDEAS PRODUCED WHEN BRAINSTORMING ARE NOT EVALUATED. EVALUATION COMES LATER WHEN ONE IS DOING DECISION MAKING AND VALUING. THERE ARE SEVERAL WAYS TO DO BRAINSTORMING.

ONE KIND OF BRAINSTORMING WHICH HELPS TO OVERCOME CONCEPTUAL BLOCKS IS ATTRIBUTE LISTING. BY LISTING ATTRIBUTES (CHARACTERISTICS) THE STUDENTS ARE HELPED TO DEVELOP THE FULLEST EXPANSION OF AN IDEA OR A PROBLEM AND THEIR THOUGHTS BEGIN TO FLOW. THIS KIND OF BRAINSTORMING CAN BE DONE INDIVIDUALLY OR BY A GROUP.

FREE WRITING IS ANOTHER APPROACH TO BRAINSTORMING WHERE THE STUDENTS WRITE CONTINUOUSLY FOR 1-5 MINUTES ON A SUBJECT. FREE WRITING ALLOWS STUDENTS TO MAKE CONNECTIONS BETWEEN THE PROBLEM AND OTHER EXPERIENCES WHICH THEY HAVE CONSCIOUSLY OR SUBCONSCIOUSLY LINKED TO THE PROBLEM. THERE IS NO EVALUATION OR EDITING OF IDEAS OR WRITING STYLE AT THIS TIME.

CLUSTERING IS AN EXERCISE WHERE A PHRASE OR WORD IS PLACED IN THE CENTER OF THE PAGE AND THE STUDENTS FOR 1-5 MINUTES JOT DOWN THE IDEAS AND WORDS THAT COME TO MIND (FREE ASSOCIATION) AND CONNECT THEM TO THE MAIN PHRASE. WHEN A SPECIFIC LINE OF FREE ASSOCIATION STOPS ANOTHER LINE IS STARTED. THE END RESULT IS A DIAGRAM OF WORDS CONNECTED TO THE MAIN PHRASE BY LINES.

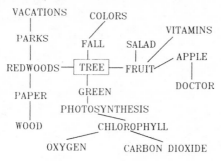

I USE THIS STRATEGY EXTENSIVELY IN CLASS FOR ANCHORING
ASSIGNMENTS, SUMMARY ASSIGNMENTS AND TO DEMONSTRATE ONE WAY TO A
APPROACH AN ESSAY QUESTION (KLOSS, 1986).

NUTSHELLING

NUTSHELLING IS STATING IN TWO OR THREE SENTENCES THE MAIN
IDEAS IN AN ASSIGNMENT AND HOW THEY ARE RELATED. IT FORCES THE
STUDENTS TO DISTINGUISH MAJOR IDEAS FROM MINOR ONES AND FOCUSES
THEM ON THE MOST IMPORTANT ISSUE IN THE ASSIGNMENT. THIS MEANS
THE STUDENTS MUST CREATE A UNIFYING IDEA OR NEW CONCEPT BY
CONDENSING, SYNTHESIZING, AND COMBINING IDEAS. I USE THIS
STRATEGY IN MY LABORATORY EVALUATION ASSIGNMENT AND IN SUMMARY
LECTURE ASSIGNMENTS.

TEACHING TO OTHERS

BY TEACHING AN IDEA OR EXPLAINING A CONCEPT OR A SOLUTION TO
A PROBLEM TO SOMEONE ELSE THE STUDENTS MUST IDENTIFY POINTS,
ORGANIZE THEIR THOUGHTS AND SYNTHESIZE INFORMATION. BY TRYING TO
TEACH OR EXPLAIN TO ANOTHER STUDENT, FLAWS IN REASONING CAN BE
SPOTTED, WEAK AREAS IDENTIFIED AND QUESTIONS RAISED ABOUT UNCLEAR
AREAS.

GRAPHIC REPRESENTATION OR ILLUSTRATION (VISUAL IMAGING)

THE STUDENTS TAKE VERBAL INFORMATION OR NUMERICAL DATA AND
FIND A WAY TO VISUALLY REPRESENT OR DEMONSTRATE IT BY DRAWING A
PICTURE, DIAGRAM, OR GRAPH. THIS STRATEGY FORCES THE USE OF
IMAGINATION, CHANGES THE NORMAL FRAMEWORK OR REFERENCE SET, AND
GIVES THE STUDENTS A NEW INSIGHT INTO THE MATERIAL.

CHAPTER 7

HOW I USE WRITING

I USE WRITING FOUR WAYS IN LECTURE.

1. LOGS (SEE APPENDIX I, ASSIGNMENT #1). LOG ASSIGNMENTS ARE USED TO ESTABLISH COMMUNICATION BETWEEN MYSELF AND THE STUDENTS.

2. ANCHORING EXERCISES. THESE ARE IN CLASS WRITING ASSIGNMENTS USUALLY DONE EITHER BY CLUSTERING OR FREE WRITING. THE STUDENTS INDIVIDUALLY INTERACT WITH THE IDEA, CONCEPT OR THEORY THAT WE ARE GOING TO BE DISCUSSING THAT SESSION. THE ANCHORING EXERCISES CAN ACT AS A BRIDGE BETWEEN THE STUDENTS' FORMER ACTIVITY AND THE CLASS. THEY ALSO CAN BE USED TO RE-ENGAGE THE STUDENTS DURING A LECTURE, TO HELP CLARIFY THEIR THINKING ABOUT AN ISSUE AND TO CHECK ON THEIR LEVEL OF UNDERSTANDING. THE EXERCISES HELP TO SHOW THEM CONNECTIONS BETWEEN WHAT THEY ALREADY KNOW AND THE CLASS MATERIAL. ANCHORING EXERCISES GIVE VALIDITY TO THEIR OWN EXPERIENTIAL KNOWLEDGE. AFTER THE EXERCISE I HAVE STUDENTS EXCHANGE THEIR CLUSTERS OR FREE WRITINGS WITH A PERSON SITTING NEXT TO THEM. THE STUDENTS THEN CIRCLE TWO ITEMS ON THEIR NEIGHBOR'S WORK THAT ARE NEW OR DIFFERENT FROM THEIR OWN. THIS INCREASES THE STUDENTS EXPOSURE TO DIFFERENT WAYS OF 'SEEING' THE TOPIC. THIS EXPOSURE TO DIFFERENT POINTS OF VIEW CAN ALLOW THEM TO SHIFT THEIR OWN FRAME OF REFERENCE. AFTER THE STUDENTS HAVE RETURNED THEIR NEIGHBOR'S WORK THERE IS A BRIEF CLASS DISCUSSION. I USUALLY PUT THEIR IDEAS ON THE BLACK BOARD AND THEN AS I LECTURE I REFER TO THE LIST. I USE ANCHORING AT THE BEGINNING OF EVERY NEW MAJOR TOPIC AND AT TIMES AS A BRIEF REVIEW OF THE PREVIOUS LECTURE.

3. SUMMATION OR CONCLUSION EXERCISES. THESE EXERCISES ARE DONE AT THE END OF A TOPIC. THEY MAY BE DONE EITHER ENTIRELY IN CLASS OR THEY MAY BE ASSIGNED AT THE END OF THE PRECEDING LECTURE SESSION AND BROUGHT TO THE FOLLOWING SESSION FOR IN CLASS PROCESSING. I DISTRIBUTE A WRITING ASSIGNMENT AND THE CLASS MAY WORK IN PEER GROUPS OR FIRST INDIVIDUALLY AND THEN IN PEER GROUPS. THE GROUPS ALWAYS SHARE THEIR WORK WITH THE WHOLE CLASS AND THIS IS FOLLOWED BY A CLASS DISCUSSION TO TIE THE WHOLE TOPIC AND CONCLUSIONS TOGETHER. THE PURPOSES OF THESE ASSIGNMENTS ARE TO HELP THE STUDENTS LEARN THE MATERIAL, TO SEE WHAT THEY DON'T UNDERSTAND, TO SEE WHERE THERE IS CONFUSION, TO PRACTICE SUMMARIZING INFORMATION, TO WORK ON OTHER ACADEMIC ABILITIES, TO EMPHASIZE CLASS OBJECTIVES AND TO IMPROVE COLLABORATIVE AND ACTIVE LEARNING.

4. ESSAY QUESTIONS. THESE ARE DONE INDIVIDUALLY OR IN A PEER GROUP IN CLASS, AS A STUDY GROUP ASSIGNMENT OR AS PART OF AN EXAM. THE FIRST TIME THE STUDENTS DO AN ESSAY QUESTION ASSIGNMENT FOR ME IT IS DONE IN CLASS. I GIVE A PRACTICE ESSAY QUESTION AND THE STUDENTS WRITE THEIR ANSWERS. THEN PEER GROUPS

ARE FORMED AND EACH STUDENT READS THEIR ANSWER TO THE GROUP.
AFTER DISCUSSING THE INDIVIDUAL ANSWERS THE GROUP SELECTS A
RESPONSE OR WRITES A GROUP ANSWER TO PRESENT TO THE CLASS. AFTER
THE CLASS PRESENTATIONS THERE IS A CLASS DISCUSSION ABOUT WHAT
THE QUESTION ASKED FOR AND HOW TO GO ABOUT ANSWERING IT. I USE A
COMPARE QUESTION AND USUALLY GET AT LEAST ONE LISTING ANSWER.
THIS LEADS INTO A DISCUSSION OF HOW TO ANSWER ESSAY QUESTIONS. I
THEN HAND OUT AND GO OVER THE ESSAY WORD LIST (SEE CHAPTER 10 AND
APPENDIX III, ASSIGNMENT #41). I ALSO HAND OUT SAMPLE QUESTIONS
WITH 'GOOD' AND 'BAD' ANSWERS FROM A PREVIOUS CLASSES SO WE CAN
DISCUSS GRADING CRITERIA AND THE STUDENTS CAN ACTUALLY HAVE
PRACTICE IN DEVISING SCORING GUIDES. THIS IS DONE IN PEER GROUPS
FOLLOWED BY CLASS DISCUSSION. AFTER THIS IN CLASS SESSION ON
ESSAY QUESTIONS I PERIODICALLY ASSIGN INDIVIDUAL HOMEWORK ESSAY
QUESTIONS WHICH ARE BROUGHT INTO CLASS AND DISCUSSED IN PEER
GROUPS. I READ THE STUDENTS' ANSWERS AND COMMENT ON THEM.
DURING THE SEMESTER THE STUDY GROUPS ARE ASSIGNED THE JOB OF
WRITING AND ANSWERING TWO ESSAY QUESTIONS BASED ON LECTURE
MATERIAL (SEE APPENDIX III, ASSIGNMENT #42). THESE ARE HANDED IN
ONE WEEK PRIOR TO EACH OF THE HOURLY EXAMS. I REPRODUCE THE
STUDY GROUPS' QUESTIONS AND ANSWERS FOR THE CLASS AND GO OVER
THEM IN CLASS. EACH HOURLY EXAM INCLUDES ESSAY QUESTIONS (SEE
CHAPTER 10 FOR SAMPLE QUESTIONS).

I USE WRITING FOUR WAYS IN LABORATORY.

1. LABORATORY JOURNALS (SEE APPENDIX I, ASSIGNMENT #2). THE
PURPOSE OF THE JOURNAL IS TO HAVE THE STUDENTS EXPERIENCE BIOLOGY
OUTSIDE OF LECTURE AND LABORATORY IN THE REAL WORLD. I WANT TO
BROADEN THE STUDENTS' PERCEPTION OF WHAT BIOLOGY IS AND HOW
OTHERS SEE IT. THIS INCREASES THE STUDENTS' AWARENESS OF
DIFFEENT MODELS OF REALITY AND SHIFTS THEIR POINT OF VIEW. I
WANT THEM TO REFLECT ON THEIR EXPERIENCES. THE LABORATORY
JOURNAL IS A PERSONAL RESPONSE TO AN EXPERIENCE AND THEREFORE
THEY CAN'T BE WRONG.

2. LABORATORY EVALUATIONS (SEE APPENDIX I, ASSIGNMENT #3). THESE
WEEKLY EVALUATIONS ARE DONE BY THE STUDENTS TO HELP ME IMPROVE
THE LABORATORY AND MY TEACHING. THE EVALUATIONS FORCE THE
STUDENTS TO ANALYZE THEIR EXPERIENCES AND THEN SUMMARIZE OR
IDENTIFY THE PURPOSE OF THE LABORATORY. THIS IS VERY DIFFICULT
FOR MANY STUDENTS AS THEY CAN'T SEE THE FOREST FOR THE TREES.

3. WEEKLY WRITING ASSIGNMENTS (SEE APPENDIX II, ASSIGNMENTS # 5-
30). THESE ASSIGNMENTS ARE TAILORED FOR EACH LABORATORY
EXPERIENCE AND SPECIFIC OBJECTIVES ARE SPELLED OUT FOR THE
STUDENTS ON THE ASSIGNMENT.

4. ESSAY QUESTIONS ON EXAMINATIONS. THE QUESTIONS ARE DESIGNED
TO LET THE STUDENTS DEMONSTRATE NOT ONLY THAT THEY HAVE LEARNED
THE CONTENT OF THE LABORATORIES BUT HAVE ALSO LEARNED TO OBSERVE,
ORGANIZE DATA, REPRESENT DATA, IDENTIFY ASSUMPTION, SUMMARIZE AND

DRAW CONCLUSIONS, MAKE INFERENCES AND PRESENT DATA GRAPHICALLY.

CHAPTER 8

SMALL GROUPS OR PEER GROUPS

IN A LARGE CLASS INDIVIDUAL STUDENTS MAY SUFFER A LOSS OF
LEARNING (MCKECHIE, 1986) BUT THIS MAY BE ALLEVIATED BY USING
SMALL (PEER) GROUPS OF FOUR OR FIVE. MANY THINGS CAN HAPPEN IN
SMALL GROUP WORK BESIDE INCREASED LEARNING. GROUP WORK CAN BUILD
TRUST (JUDY AND JUDY, 1981), DECREASE MISUNDERSTANDING OF THE
MATERIAL, DEMONSTRATE EMOTIONAL BIASES, RE-ORGANIZE LEARNING AND
EXPERIENCE, DEVELOP A SENSE OF COMMUNITY, AND INCREASE THE
STUDENTS' SENSE OF RESPONSIBILITY. STUDENTS ACTUALLY LEARN
MATERIAL BY TEACHING (EXPLAINING) IT TO THEIR PEERS. IN SMALL
GROUP WORK STUDENTS TALK TO OTHER STUDENTS (NON-SPECIALISTS) AS
AN AUDIENCE AND THEREFORE ARE FORCED TO TRANSLATE SCIENCE SO THAT
IT IS UNDERSTANDABLE. BY READING THEIR WORK ALOUD TO THE GROUP
THE STUDENTS RECEIVE IMMEDIATE RESPONSE TO THE CONTENT (WHITE,
1986) AND THE READER IS SEEN AS A REAL PERSON. STUDENTS FIND OUT
IF THEY ARE REALLY COMMUNICATING AND THEN CAN RE-WRITE DRAFT
MATERIAL WITH THEIR CLASSMATES' RESPONSES IN MIND. BY WORKING IN
A GROUP THE STUDENTS GAIN EXPERIENCE IN EVALUATING AND EDITING IN
A LESS HOSTILE ATMOSPHERE (JUDY AND JUDY, 1981).

MY CLASSES FREQUENTLY WORK IN SMALL GROUPS FOR ANCHORING AND
SUMMARIZING EXERCISES. I ALSO ORGANIZE STUDY GROUPS THE FIRST
DAY OF CLASS (SEE APPENDIX III, ASSIGNMENT #42). THE FIRST TIME
THE CLASS DOES GROUP WORK WE DISCUSS THE KINDS OF COMMENTS THE
STUDENTS CAN MAKE. THE FIRST COMMENTS THEY MAKE ON ANY
PRESENTATION SHOULD BE POSITIVE E.G. WHAT DID YOU LIKE BEST OR
WHAT WAS THE MOST SUCCESSFUL PART OF THE WORK. POSITIVE COMMENTS
ARE FOLLOWED BY ONE OR TWO COMMENTS ON WHAT COULD BE IMPROVED OR
CLARIFIED. DID YOU UNDERSTAND IT, WAS IT CLEAR, WAS IT ACCURATE.
I ALSO MAKE IT CLEAR THAT EACH GROUP MEMBER HAS A RESPONSIBILITY
TO THE GROUP TO HAVE WORK TO DISCUSS. AFTER A PRESENTOR HAS READ
OR SHOWN (VISUAL REPRESENTATION) THEIR WORK, THEY RECORD THE
GROUP'S RESPONSES ON THE BACK OF THE PAPER. IF THE GROUP
COMMENTS ARE NOT RECORDED THEY SEEM TO SLIP AWAY AND HAVE MUCH
LESS WEIGHT WHEN IT COMES TO DISCUSSING OR REVISING THE WORK. TO
HELP THEM KEEP TRACK OF WHAT THEY ARE SUPPOSED TO BE FOCUSING ON
YOU CAN HAND OUT WORKSHEETS OR PUT THE INFORMATION ON THE BOARD.

CHAPTER 9

ASSESSMENT AND EVALUATION

As you will see in my course assignments my classes do a lot of writing and the question arises how do I assess or evaluate it all? What about the time demands of adding writing to your courses? Now that I incorporate writing in my class I do spend more time in preparation but most of that time is in the planning or designing of writing assignments and the course itself not in grading papers. Although I do a lot of assessing or evaluating it takes less time than you might think. With all this writing going on the question is how can I (the teacher) grade (asses, evaluate) it all? Where do I get the time, energy and at times the enthusiasm? The answer is that different assignments are evaluated or assessed differently and some not at all.

What kinds of evaluation or assessment are available?
1. teacher or authority evaluation
2. peer critiquing
3. self-critiquing

Not all writing requires a teacher's response. This was the hardest lesson for me to learn and sometimes for the students to accept. Not all of the writing I do evaluate receives a grade. There are many ways to evaluate or assess work and I use different methods for different assignments.

Types of evaluation:
GRADE A TO F
PASS/FAIL
CHECK - MEANS IT WAS DONE CAN BE USED WITH OR WITHOUT A
+/- OPTION
SCALE - EVALUATING USING A NUMBER SCALE CAN BE USED
WITH OR WITHOUT A +/- OPTION
EXAMPLE OF A SCALE:
3- DID THE ASSIGNED TASK, SERIOUS EFFORT
2- DID THE ASSIGNED TASK, WORK HASTY, BRIEF,
LITTLE THOUGHT
1- CURSORY WORK
0- WORK NOT DONE
COMMENTS WITH OR WITHOUT AN EVALUATION - COMMENTS CAN BE
FOCUSED ON JUST MAJOR STRENGTHS AND ONE PROBLEM
AREA THAT NEEDS WORK. COMMENTS ARE USED TO
ENCOURAGE THINKING.
INDICATION MARKS - NO WRITTEN COMMENTS ARE MADE BUT
MARKS ARE USED E.G. UNDERLINE THE CENTRAL IDEA,
CHECK GOOD POINT(S), QUESTION MARKS FOR UNCLEAR
AREAS.

I evaluate or assess for content NOT for correct English. I do however at the beginning of each semester talk to the students

ABOUT THE NEED TO USE GOOD ENGLISH FOR COMMUNICATION. I DO TAKE HALF CREDIT OFF ON EXAMS FOR MISSPELLED SCIENTIFIC WORDS. IF STUDENTS HAVE REAL COMMUNICATION PROBLEMS I REFER THEM TO THE CAMPUS TUTORING CENTER.

I HAVE RECENTLY READ EDWARD WHITE'S (1986) BOOK WHICH COVERS THE THEORIES, ARGUMENTS AND PRACTICES ON ASSESSMENT. EVEN THOUGH THIS BOOK IS ABOUT WRITING, THE MAJOR CONCEPTS, POINTS AND IDEAS ARE CERTAINLY APPLICABLE TO ANY AREA. INSIGHTS INTO HOW WE ASSESS OUR COURSES CAN IMPROVE OUR TEACHING AND VICE VERSA. WHEN I ASK MYSELF WHAT OBJECTIVES AM I TESTING TOWARD? I CAN THEN LOOK AT WHAT I TEACH AND HOW I TEACH TO SEE IF I AM ACCOMPLISHING WHAT I WANT. I CAN ALSO LOOK AT THE ASSESSMENT TOOL AND ASK IF IT IS MEASURING WHAT I AM ACTUALLY DOING AND WANT TO DO IN THE CLASS. THE USE OF SCORING GUIDES AND GRADING CRITERIA RELATE TEACHING TO ASSESSMENT AND SUPPORT GOOD TEACHING.

IT IS VERY IMPORTANT TO SHARE THE CRITERIA FOR EVALUATION OR ASSESSMENT WITH THE STUDENTS. THE GRADING CRITERIA MUST BE CLEAR. WHEN I CLARIFY GRADING CRITERIA I OFTEN FIND THAT I AM NOT ASSESSING WHAT I WANT TO OR THOUGHT I WAS. MUCH TEST ANXIETY FOR THE STUDENTS COMES FROM UNCLEAR GRADING CRITERIA OR UNCLEAR TESTS. STUDENTS SHOULD NOT HAVE TO GUESS AT WHAT IS BEING ASKED, THEY NEED CLEAR DIRECTIONS.

WHEN HANDING OUT ASSIGNMENTS THEY SHOULD INCLUDE THE GRADING CRITERIA AND WHEN NECESSARY A SCORING GUIDE. THE SCORING GUIDE CAN BE MADE UP BY YOU ALONE OR WITH THE STUDENTS. HOWEVER, GOING THROUGH THE GRADING CRITERIA AND SCORING GUIDE WITH THE STUDENTS IS IMPORTANT. IF SAMPLE PAPERS ARE AVAILABLE THESE CAN BE GIVEN OUT SO THAT THE STUDENTS CAN EVALUATE THEM USING THE GRADING CRITERIA AND SCORING GUIDE. THIS KIND OF EXPERIENCE HELPS THE STUDENTS UNDERSTAND HOW THE SCORING GUIDE IS APPLIED. I DO THIS WORK IN GROUPS FOR ESSAY EXAMS AND OTHER ASSIGNMENTS.

CHAPTER 10

EXAMS

What Kind Do I Give and Why?

My first teaching experience was as a graduate assistant in a 30 section General Biology course. The lecture, presented on closed circuit television to classes of 24 students, was followed by a short recitation lead by the teaching graduate assistants. The common exams which were multiple choice with at least one essay were graded by the teaching assistants. Needless to say when I started to teach my own courses the exams were multiple choice with one or two essays. No more - now my exams are definition, fill in the blank with one or two sentence answers and essay. Why the change? I discovered that with multiple choice exams I had a couple of options, I could either use test questions that came with the text with little or no modification or make up my own questions from scratch. Neither alternative worked for me. If I used the test questions that the publishers supplied the phrasing wasn't mine and the students were confused by it. If I made up the questions (which wasn't easy) the questions were either too easy or too hard or too unclear (ambiguous). Students found ways to interpret questions which when we discussed their answers were certainly legitimate. However, since not all students will debate an answer with a teacher, I felt that some students were being placed at a disadvantage. Another problem with multiple choice questions was that I wasn't sure if the students were selecting the correct answers because they knew the information or just recognized the right answer from familiarity with the material. Also, the multiple choice exams couldn't demonstrate how the students were processing the material. All in all it was very frustrating. So I changed the format. Gone was the fast grading machine. Problems arose with the new format, at first my questions were not stated clearly enough for the students to know what I wanted and the exams took more time to grade. The problem of lack of clarity has diminished but it DOES take time to grade a non-multiple choice exam.

One of the major problems with giving essay exams is that the students don't know what the instruction words mean or how to approach the question. Prior to the first exam I work with the students on how to decide what a question is asking and what is needed to answer it. I now make it clear what I want the student to do. We go over the instruction words in class and I give assignments so that the students get practice writing and answering their own essay questions as well as ones I set for them. I give them a list of words that they need to understand which are often used in essay questions. I also work with the students ahead of time on grading criteria and scoring guides.

SINCE ESSAY QUESTIONS ARE NOT JUST A LIST OF FACTS OR REGURGITATION OF INFORMATION THEY REQUIRE THAT STUDENTS SYNTHESIZE AND INTEGRATE INFORMATION. THEY DEMAND SOME LEVEL OF CRITICAL THINKING.

ESSAY INSTRUCTION WORDS:

EXPLAIN - TELL HOW IT WORKS
COMPARE - SHOW HOW THEY ARE SIMILAR AND HOW THEY ARE DIFFERENT
CONTRAST - SHOW HOW THEY ARE DIFFERENT
DESCRIBE - GIVE A DETAILED ACCOUNT
ANALYZE - DIVIDE INTO PARTS, EXPLAIN EACH PART AND HOW THEY DIFFER
SUMMARIZE - STATE BRIEFLY, GIVE A CONDENSED STATEMENT, PRESENT THE SUBSTANCE OF AN IDEA
EVALUATE - DETERMINE OR FIND THE VALUE OF, TO APPRAISE
DEFINE - TO STATE OR EXPLAIN THE MEANING OF
DISCUSS - TO WRITE ABOUT, CONSIDER AND ARGUE THE PROS AND CONS
PROVE - TO ESTABLISH A POINT BY USING EVIDENCE
ILLUSTRATE - TO MAKE CLEAR OR EASILY UNDERSTANDABLE BY USING EXAMPLES
INTERPRET - TO EXPLAIN THE MEANING OF, TO SHOW ONE UNDERSTANDING OF THE MEANING

WHEN I USE ONE OF THESE WORDS IN AN ESSAY QUESTION I PUT A TRANSLATION IN PARENTHESES FOR THE STUDENTS E.G. COMPARE MITOSIS AND MEIOSIS (TELL HOW THE PROCESSES ARE SIMILAR AND HOW THEY ARE DIFFERENT FROM EACH OTHER - THIS IS NOT A LIST).

SAMPLE ESSAY QUESTIONS

THE FOLLOWING ARE SAMPLE ESSAY QUESTIONS I HAVE USED IN GENERAL BIOLOGY I AND II LECTURE EXAMINATIONS.

1. GRAPH THE FOLLOWING REACTION AND THEN INTERPRET YOUR GRAPH (EXPLAIN THE MEANING OF THE GRAPH TO SHOW THAT YOU UNDERSTAND WHAT IT MEANS). BE SURE TO LABEL BOTH AXES OF THE GRAPH FULLY.

T + S --> P   CHANGE IN FREE ENERGY (G) = -22.6 KCAL/MOL

2. DESCRIBE (GIVE A DETAILED ACCOUNT OF) THE CONDITIONS THAT WERE NECESSARY FOR OPARIN'S THEORY OF THE ORIGIN OF LIFE TO WORK AND THEN EXPLAIN WHY HIS THEORY WOULDN'T WORK TODAY.

3. USING AN EXAMPLE OF YOUR CHOICE EXPLAIN (TELL HOW IT WORKS) THE THEORY OF EVOLUTION AS POSTULATED BY DARWIN. UNDERLINE ALL THE PROCESS WORDS.

4. COMPARE (SHOW HOW THEY ARE SIMILAR AND HOW THEY ARE DIFFERENT) THE MEMBRANE MODELS OF DAVSON-DANIELLI AND SINGER-NICHOLSON. HOW IS THE SINGER-NICHOLSON MODEL USEFUL IN EXPLAINING (TELLING HOW

IT WORKS) HOW DIFFERENT MEMBRANES ARE UNIQUE?

5. EXPLAIN (TELL HOW IT WORKS) THE MITCHELL HYPOTHESIS OF CHEMIOSMOTIC POSPHORYLATION IN TERMS OF THE STRUCTURE THE CHLOROPLAST AND THE PROCESS OF PHOTOSYNTHESIS. BE SURE TO EXPLAIN HOW PLASTOQUIONE FITS INTO THE HYPOTHESIS AND HOW IT FUNCTIONS.

6. DISCUSS (WRITE ABOUT) HOW THE THREE PARTS OF CELLULAR RESPIRATION RELATE TO EACH OTHER IN TERMS OF THE ENTIRE PROCESS AND THE STRUCTURE OF THE ORGANELLE.

7. COMPARE (SIMILARITIES AND DIFFERENCES) THE PROCESS OF MITOSIS AND MEIOSIS. WHY OR HOW IS EACH PROCESS IMPORTANT TO THE EXISTENCE OF AN ORGANISM AND/OR A SPECIES?

8. EXPLAIN (TELL HOW IT WORKS) THE MITCHELL HYPOTHESIS IN TERMS OF THE ELECTRON TRANSPORT SYSTEM. INCLUDE A FULLY LABELED DIAGRAM.

9. DISCUSS (WRITE ABOUT, CONSIDER AND ARGUE THE PROS AND CONS) IN DETAIL HOW A GENIC DELETION MIGHT EFFECT THE PHENOTYPE OF AN INDIVIDUAL. YOUR DISCUSSION MUST INCLUDE THE MECHANISM FOR PROTEIN SYNTHESIS.

10. EXPLAIN (TELL HOW IT WORKS) THE RELATIONSHIP BETWEEN HEMOGLOBIN, OXYGEN, CARBON DIOXIDE, PARTIAL PRESSURE AND ACIDITY IN HUMAN GAS EXCHANGE.

11. MAINTAINING A CONSTANT LEVEL OF BLOOD GLUCOSE IS VERY IMPORTANT IN MAN, DESCRIBE (GIVE A DETAILED ACCOUNT) HOW YOUR BLOOD SUGAR LEVEL WOULD BE RETURNED TO NORMAL IF YOU ATE A POUND OF CANDY.

12. THE DIGESTIVE PROCESS IS VERY COMPLEX. USING THE HUMAN DIGESTIVE PROCESS, EXPLAIN (TELL HOW IT WORKS) THE PROCESS OF DIGESTING AN EGG MCMUFFIN.

13. EXPLAIN (TELL HOW IT WORKS) HOW A BLOOD CLOT IS FORMED. EVALUATE (DETERMINE THE VALUE OF) THE IMPORTANCE OF CALCIUM TO THE PROCESS.

14. EXPLAIN (TELL HOW IT WORKS) THE TRANSPIRATION-COHESION-TENSION THEORY.

15. EVALUATE (DETERMINE THE VALUE OF, APPRAISE) THE IMPORTANCE OF AN INCREASE IN THE HYDROSTATIC PRESSURE IN THE EXCHANGE OF MATERIALS IN A CAPILLARY BED.

16. FRESH WATER FISH HAVE PROBLEMS MAINTAINING THE PROPER SALT AND WATER BALANCE, EXPLAIN (TELL HOW IT WORKS) HOW THE FISH DEALS WITH THESE PROBLEMS.

17. THE MAMMALIAN KIDNEY IS A WONDERFUL EXAMPLE OF HOW AN ORGANISM, USING FEEDBACK MECHANISMS, CAN MAINTAIN HOMEOSTASIS. DISCUSS (WRITE ABOUT) HOW THE FUNCTIONAL UNIT OF THE KIDNEY WORKS TO PRODUCE HIGHLY CONCENTRATED URINE. (BE SURE TO INCLUDE STRUCTURAL/FUNCTIONAL RELATIONSHIPS AND THE NECESSARY CONTROL MECHANISMS).

18. USING THE THYROID GLAND AS YOUR EXAMPLE, EXPLAIN (TELL HOW IT WORKS) HORMONAL CONTROL OF THE METABOLIC RATE IN MAN.

19. DESCRIBE (GIVE A DETAILED ACCOUNT OF) HOW THE TWO MESSENGER MODEL OF HORMONE ACTION WORKS.

20. EXPLAIN (TELL HOW IT WORKS) THE THEORIES THAT EXPLAIN FLOWERING.

21. EXPLAIN (TELL HOW IT WORKS) HOW INFORMATION FROM THE ENVIRONMENT IS TRANSMITTED TO AN EFFECTOR CELL. (BE SPECIFIC IN TERMS OF VOCABULARY AND THE MECHANISMS INVOLVED).

22. DRAW AN S SHAPED GROWTH CURVE. LABEL ITS STAGES AND SUMMARIZE (STATE BRIEFLY) WHAT THE GRAPH TELLS YOU ABOUT THE POPULATION'S CHARACTERISTICS AND HOW THE POPULATION REGULATES ITS SIZE.

23. ANALYZE (DIVIDED INTO PARTS, EXPLAIN EACH PART AND HOW THEY DIFFER) THE ASSUMPTIONS OF THE HARDY-WEINBERG LAW. DISCUSS HOW THE LAW RELATES TO THE SELECTIONIST THEORY OF EVOLUTION.

24. DESCRIBE (GIVE A DETAILED ACCOUNT OF) THE MOLECULAR BASIS OF MUSCLE CONTRACTION, EXPLAIN (TELL HOW) IT WORKS AND HOW IT IS CONTROLLED. ILLUSTRATE (MAKE CLEAR USING AN EXAMPLE) HOW KNOWLEDGE OF MUSCLE CONTRACTION AND CONTROL COULD BE USED IN INSECT CONTROL PROGRAMS.

CHAPTER 11

COURSE PLANNING INFORMATION

I USED THE FOLLOWING COURSE PLANNING WORK SHEET TO REMIND ME OF ALL THE IDEAS, THEORIES, ACADEMIC ABILITIES, AND SPECIFICS I WANTED TO THINK ABOUT AS I PLANNED GENERAL BIOLOGY I AND II.

PLANNING WORK SHEET

COURSE TILE:

CREDITS:

LECTURE: LABORATORY:

PREREQUISITES:

DAYS: TIME: ROOM:

LECTURER: NAME   OFFICE   PHONE NUMBER   OFFICE HOURS

POPULATION: MAJOR AND/OR NON-MAJOR
            WHAT LEVEL: FRESHMAN, SOPHOMORE, JUNIOR, SENIOR,
                        GRADUATE
            PART TIME AND/OR FULL TIME

SCHEDULING: DAY AND/OR NIGHT
            WILL I BE TEACHING BOTH THE LECTURE AND LAB?

CLASS SIZE: LECTURE AND LABORATORY

TEXT: LECTURE 1
                 2 ETC
        LABORATORY 1
                   2 ETC

TEACHING METHODOLOGIES: WHAT PROCESS METHODOLOGIES DO I WANT TO
                        USE?

GOALS AND OBJECTIVES:
    CONTENT: WHAT RECURRENT THEMES DO I WANT TO USE?
             WHAT SPECIFIC BIOLOGICAL SKILLS AND TECHNIQUES DO I
             WANT TO EMPHASIZE IN THIS COURSE?
    STUDENT COGNITIVE DEVELOPMENT:
        LEARNING STYLES: HOW WILL I INCORPORATE LEARNING STYLES
                         AND DIFFERENT WAYS OF KNOWING IN THE
                         COURSE?
        ACTIVE LEARNER: HOW WILL I STRUCTURE THE COURSE TO MAKE
                        THE STUDENTS ACTIVE LEARNERS?
        WAYS OF LEARNING AND KNOWING: HOW WILL I STRUCTURE THE
                                      COURSE SO THAT THE STUDENTS

ARE EXPOSED TO ALL THE WAYS
OF LEARNING AND KNOWING?

PERRY SCHEMA: WHERE ARE THE STUDENTS ON THE PERRY SCHEMA?
WHICH STAGE? WHAT ARE THE CHARACTERISTICS
OF THAT STAGE? WHAT CHALLENGES CAN BE USED
TO MOVE THE STUDENTS TOWARDS THE NEXT
STAGE?

LANGUAGE AND USAGE: ARE THE STUDENTS PRE-SOCIALIZED OR
SOCIALIZED OR POST-SOCIALIZED? HOW
WILL I STRUCTURE THE COURSE TO
INCREASE THEIR SOCIALIZATION?

WOMEN'S WAYS OF KNOWING: HOW I CAN STRUCTURE THE CLASS TO
MOVE THE STUDENTS TOWARD
CONNECTEDNESS?

CRITICAL THINKING: WHAT LEVEL ARE THE STUDENT AT? WHAT
PROCESSES OF CRITICAL THINKING CAN BE
PRACTICED IN THIS CLASS?

ACADEMIC ABILITIES: HOW CAN I STRUCTURE THE COURSE TO
IMPROVE THE STUDENTS' SKILLS?
COMMUNICATION - ORAL/WRITTEN
ANALYTICAL - EVALUATION - CRITICAL THINKING
PROBLEM SOLVING
VALUE FORMATION
SOCIAL INTERACTION
SYNTHESIS - INTEGRATION - SUMMARIZE INFORMATION
AESTHETIC APPRECIATION
AWARENESS OF OTHER CULTURES - SEEING FROM ANOTHER
PERSPECTIVE
ORGANIZATION - REFERENCING
CREATIVITY - IMAGINATION
DECISION MAKING
INTERPRETATION

WRITING: HOW AM I GOING TO USE WRITING AND PROCESS IN
CLASS? WHAT STRATEGIES AM I GOING TO USE?

ASSESSMENT AND EVALUATION: WHAT DO I WANT TO ASSESS AND/OR
EVALUATED? HOW DO I WANT TO DO IT?

CHAPTER 12

THE COURSE

BACKGROUND

THE FOLLOWING IS THE BACKGROUND INFORMATION FOR THE COURSE GENERAL BIOLOGY I AND II. THIS INFORMATION INCLUDES BASIC INFORMATION ON THE POPULATION TO BE SERVED AND THE METHODOLOGIES, GOALS, OBJECTIVES AND THEMES USED TO STRUCTURE THE COURSE.

TITLE: GENERAL BIOLOGY I AND II

PREREQUISITES: GENERAL BIOLOGY I = NONE
GENERAL BIOLOGY II = GENERAL BIOLOGY I
ALTHOUGH THE CATALOGUE STATES THAT GENERAL BIOLOGY I IS THE PREREQUISITE COURSE FOR GENERAL BIOLOGY II THIS IS NOT STRICTLY ENFORCED.

POPULATION: THE COURSE IS DESIGNED FOR SCIENCE MAJORS (BIOLOGY, CHEMISTRY, ENVIRONMENTAL STUDIES AND MATHEMATICS). STUDENTS ARE PRIMARILY FRESHMAN AND SOPHOMORES (18-20 YEARS), ALTHOUGH THERE ARE OLDER RETURNING STUDENTS IN BOTH THE DAY AND NIGHT SECTIONS. THE CLASS IS COMPOSED OF BOTH FULL AND PART-TIME STUDENTS. THE MAJORITY OF THE STUDENTS ARE COMMUTERS AND MOST HAVE JOBS EITHER ON OR OFF CAMPUS AND WORK FROM 5-40+ HOURS A WEEK.

CLASS SIZE: NORMALLY THE SIZE OF THE LECTURE RANGES FROM 24 TO 75 WITH A MINIMUM OF TEN. EACH LABORATORY SECTION CAN CONTAIN A MAXIMUM OF 24 ALTHOUGH SOME SECTIONS HAVE BEEN AS SMALL AS SIX.

SCHEDULING: THE CLASS MEETS THREE TIMES A WEEK; TWO 75 MINUTES LECTURES AND ONE 2 1/2 HOUR LABORATORY. SEVERAL SECTIONS OF THE COURSE ARE GIVEN EACH SEMESTER. ONE SECTION IS ALWAYS GIVEN AT NIGHT AND MEETS TWICE A WEEK FOR 2 1/2 HOURS EACH SESSION. DUE TO SCHEDULING CONFLICTS OFTEN I DO NOT TEACH ALL OR ANY OF MY LECTURE SECTION STUDENTS IN LABORATORY. THIS CAUSES DIFFICULTY IN ACCOMPLISHING ALL OF MY GOALS FOR THE CLASS AS OTHER FACULTY HAVE DIFFERENT APPROACHES TO THE MATERIAL AND HAVE DIFFERENT METHODS OF TEACHING. ALTHOUGH THE SYLLABUS AND LABORATORY EXERCISES ARE THE SAME FOR ALL SECTIONS OF THE COURSE, THE ACTUAL COVERAGE AND PRESENTATION OF THE MATERIAL MAY VARY GREATLY. WHAT I AM PRESENTING HERE IS HOW I TEACH THE COURSE UNDER WHAT I CONSIDER TO BE IDEAL CONDITIONS I.E. WHERE I TEACH ALL THE STUDENTS IN BOTH LECTURE AND LABORATORY, INCORPORATING CONTENT AND PROCESS TO INCREASE THE STUDENTS' LEARNING AND COGNITIVE DEVELOPMENT.

TEACHING METHODOLOGIES: I USE A VARIETY OF METHODS IN CLASS; LECTURE, CLASS DISCUSSION, INDIVIDUAL/GROUP WRITING AND DISCUSSION, STUDY GROUPS, LAB PARTNERSHIPS/TEAMS AND SEVERAL

TYPES OF WRITING ACTIVITIES. I TRY TO CREATE AN ENVIRONMENT WHICH IS NURTURING, SUPPORTIVE AND NON-COMPETITIVE. I TRY TO ESTABLISH A SENSE OF COMMUNITY THROUGH PEER GROUPS, STUDY GROUPS, LAB PARTNERSHIPS, LAB TEAMS AND THE USE OF THE LOG.

GOALS AND OBJECTIVES: THE GOALS AND OBJECTIVES FOR THIS COURSE ARE NUMEROUS. IN TERMS OF CONTENT THE GOAL IS TO ESTABLISH A COMMON BASE OF BIOLOGICAL KNOWLEDGE (VOCABULARY, CONCEPTS AND THEORIES) FOR THE STUDENTS TO BUILD ON IN LATER COURSES. THIS COURSE GIVES THE MAJORS A SIMILAR BACKGROUND SO THAT THE OTHER DEPARTMENTAL FACULTY CAN ASSUME A COMMON LEVEL OF SOCIALIZATION. THERE ARE SEVERAL RECURRENT THEMES THAT RUN THROUGH THE COURSE AND I STRESS AND TIE THEM IN WITH THE LECTURE AND LABORATORY MATERIAL AT EVERY OPPORTUNITY.

RECURRENT THEMES:
THE THEORY OF EVOLUTION AND ITS IMPORTANCE TO BIOLOGY
THE IDEA OF HIERARCHICAL STRUCTURE AND THE
INTERCONNECTIONS BETWEEN LEVELS OF COMPLEXITY
THE IDEA THAT THE WHOLE IS GREATER THAN THE SUM OF ITS PARTS
THAT BIOLOGY IS A SYNTHETIC SCIENCE
HISTORICAL PERSPECTIVE OF SCIENCE
THE RELATIONSHIP BETWEEN STRUCTURE AND FUNCTION
FEEDBACK MECHANISMS AND CONTROL
THE EXCITEMENT AND WONDER OF NATURE AND SCIENCE
THE IMPORTANCE OF THE QUESTIONING APPROACH
BIOLOGY AS STATE OF THE ART
THE IMPORTANCE OF SITUATIONAL ETHICS.

I STRUCTURE THE COURSE SO THAT THE STUDENTS NOT ONLY LEARN CONTENT BUT CAN ENHANCE THEIR ACADEMIC ABILITIES AND COGNITIVE DEVELOPMENT.

THERE ARE SEVERAL OTHER OBJECTIVES AND GOALS WHICH ARE EMPHASIZED IN THE LABORATORY. THE STUDENTS LEARN HOW TO: DESIGN EXPERIMENTS, FORM AND TEST AN HYPOTHESIS, OBSERVE AND RECORD OBSERVATIONS, IDENTIFY ASSUMPTIONS, MAKE INFERENCES, MANIPULATE DATA AND PRESENT IT IN GRAPHICAL REPRESENTATION, WORK ALONE AND WITH A PARTNER, RELATE BIOLOGY TO OTHER PARTS OF THEIR LIVES, HAVE A BROADER PERSPECTIVE, AND TO USE THEIR IMAGINATIONS.

THE SYLLABUS FOR GENERAL BIOLOGY I

GENERAL BIOLOGY I   BIO. 163   FALL

CREDITS: 4   LECTURE: DAYS _____   TIME:_____
                          ROOM:_____
              LABORATORY:DAY:_____   TIME: _____
                          ROOM:_____

LECTURER: NAME _____ OFFICE _____ PHONE _____

OFFICE HOURS _____

LABORATORY INSTRUCTOR: NAME _____ OFFICE _____
PHONE _____
OFFICE HOURS _____

WEEK DATE	SUBJECT	TEXT	LABORATORY
1	INTRODUCTION ORIGIN OF LIFE		INTRODUCTION PECHENIK CHAP. 1
2	EVOLUTION MATTER AND ENERGY		#8 STRUCTURE OF CELLS PECHENIK CHAP. 2
3	MATTER AND ENERGY CELL CHEMISTRY		MICROSCOPE HANDOUT PECHENIK CHAP. 9
4	CELL CHEMISTRY		#2 PHYSICAL PROCESSES
5	EXAM I CELL STRUCTURE AND FUNCTION		#3 SACCAHARIDES
6	CELL STRUCTURE AND FUNCTION		#4 AMINO ACIDS AND PROTEIN SYNTHESIS
7	PHOTOSYNTHESIS		LAB. MID-TERM
8	PHOTOSYNTHESIS		#5 ENZYMES
9	EXAM II RESPIRATION		#6 PHOTOSYNTHESIS
10	RESPIRATION		#7 RESPIRATION
11	CELL REPRODUCTION		#7 RESPIRATION
12	EXAM III NATURE OF THE GENE		#9 MITOSIS AND MEIOSIS
13	GENE ACTION		#10 MENDELIAN GENETICS
14	PATTERNS OF INHERITANCE		#13, 12 HUMAN AND POPULATION GENETICS
15	GENETICS		LAB. FINAL
16	FINAL		LAB. BOOK DUE

LECTURE TEXT: 1. TITLE, AUTHOR, EDITION, DATE, PUBLISHER. REQUIRED

2. ALL THE STRANGE HOURS, LOREN EISELEY, 1985, CHARLES SCHRIBNER'S SONS, NEW YORK. NOT REQUIRED, ON RESERVE IN LIBRARY.

LABORATORY TEXT: 1. EXERCISES IN BIOLOGY, HEIDMANNM, 1984, PWS-WILLARD GRANT PRESS. REQUIRED
2. A SHORT GUIDE TO WRITING ABOUT BIOLOGY, J.A. PECHENIK, 1987, LITTLE BROWN. REQUIRED.

GRADE DETERMINATION:

LECTURE: THE LECTURE COMPONENT OF THE COURSE WILL COUNT FOR 3/4 OF THE TOTAL COURSE GRADE. THE LECTURE GRADE WILL BE BASED ON YOUR TWO HIGHEST HOURLY EXAM GRADES PLUS THE CUMULATIVE FINAL. I WILL DROP THE LOWEST OF THE FIRST THREE HOURLY EXAMS AND COUNT ONLY THE HIGHEST TWO. THE FINAL EXAM MUST BE TAKEN. THE FINAL EXAM IS CUMULATIVE: 50% OLD MATERIAL (FROM THE FIRST THREE HOURLY EXAMS) AND 50% NEW MATERIAL (MATERIAL COVERED SINCE THE THIRD HOURLY EXAM). PART OF THE LECTURE GRADE WILL BE BASED ON YOUR HOMEWORK ASSIGNMENTS, GROUP WORK, STUDY GROUP ASSIGNMENTS, AND CLASS DISCUSSION.

PART OF THE LECTURE EVALUATION MAY BE THE LOGS.

LOGS: LOGS ARE DUE AT THE BEGINNING OF THE SECOND LECTURE SESSION OF THE WEEK. LATE LOGS WILL NOT BE ACCEPTED. SEE HANDOUT (SEE APPENDIX I, ASSIGNMENT #1).

LABORATORY: THE LABORATORY COMPONENT OF THE COURSE WILL COUNT FOR 1/4 OF THE TOTAL COURSE GRADE. YOU MUST PASS LAB TO PASS THE COURSE. AN "F" IN LAB MEANS AN "F" IN THE COURSE. THE LABORATORY GRADE WILL BE BASED ON TWO LAB EXAMS (MID-TERM AND FINAL), LAB NOTEBOOK, JOURNALS, AND ANY OTHER ASSIGNMENTS MADE BY YOUR INSTRUCTOR. SEE LABORATORY HANDOUTS FOR MORE INFORMATION, CLARIFICATION AND DETAIL.

EXAM POLICY: THERE WILL BE NO MAKE-UP EXAMS AND NO EXTRA CREDIT. CREDIT WILL BE LOST ON ANY QUESTION WHERE SCIENTIFIC WORDS ARE MISSPELLED. SPELLING COUNTS.

TEACHING METHODOLOGIES: I USE A VARIETY OF METHODS IN CLASS; LECTURE, CLASS DISCUSSION, INDIVIDUAL/GROUP WRITING AND DISCUSSION, STUDY GROUPS, LAB PARTNERSHIPS/TEAMS AND SEVERAL TYPES OF WRITING ACTIVITIES. I TRY TO CREATE AN ENVIRONMENT WHICH IS NURTURING, SUPPORTIVE AND NON-COMPETITIVE. I TRY TO ESTABLISH A SENSE OF COMMUNITY THROUGH PEER GROUPS, STUDY GROUPS, LAB PARTNERSHIPS, LAB TEAMS AND THE USE OF THE LOG.

GOALS AND OBJECTIVES: THE GOALS AND OBJECTIVES FOR THIS COURSE ARE NUMEROUS. IN TERMS OF CONTENT THE GOAL IS TO ESTABLISH A COMMON BASE OF BIOLOGICAL KNOWLEDGE (VOCABULARY,

CONCEPTS AND THEORIES) FOR THE STUDENTS TO BUILD ON IN LATER COURSES. THIS COURSE GIVES THE SCIENCE MAJORS A SIMILAR BACKGROUND SO THAT OTHER DEPARTMENTAL FACULTY CAN ASSUME A COMMON LEVEL OF SOCIALIZATION. THERE ARE SEVERAL RECURRENT THEMES THAT RUN THROUGH THE COURSE AND I STRESS AND TIE THEM IN WITH THE LECTURE MATERIAL AT EVERY OPPORTUNITY.

RECURRENT THEMES:
THE THEORY OF EVOLUTION AND ITS IMPORTANCE TO BIOLOGY
THE IDEA OF HIERARCHICAL STRUCTURE AND THE
INTERCONNECTIONS BETWEEN LEVELS OF COMPLEXITY
THE IDEA THAT THE WHOLE IS GREATER THAN THE SUM OF ITS
PARTS
THAT BIOLOGY IS A SYNTHETIC SCIENCE
HISTORICAL PERSPECTIVE OF SCIENCE
THE RELATIONSHIP BETWEEN STRUCTURE AND FUNCTION
FEEDBACK MECHANISMS AND CONTROL
THE EXCITEMENT AND WONDER OF NATURE AND SCIENCE
THE IMPORTANCE OF THE QUESTIONING APPROACH
BIOLOGY AS STATE OF THE ART
THE IMPORTANCE OF SITUATIONAL ETHICS.

I STRUCTURE THE COURSE SO THAT THE STUDENTS NOT ONLY LEARN CONTENT BUT CAN INCREASE THEIR ACADEMIC ABILITIES AND COGNITIVE DEVELOPMENT.

THERE ARE SEVERAL OTHER OBJECTIVES AND GOALS WHICH ARE EMPHASIZED IN THE LABORATORY. THE STUDENTS LEARN HOW TO: DESIGN EXPERIMENTS, FORM AND TEST AN HYPOTHESIS, OBSERVE AND RECORD OBSERVATIONS, IDENTIFY ASSUMPTIONS, MAKE INFERENCES, MANIPULATE DATA AND PRESENT IT IN GRAPHICAL REPRESENTATION, WORK ALONE AND WITH A PARTNER, RELATE BIOLOGY TO OTHER PARTS OF THEIR LIVES, HAVE A BROADER PERSPECTIVE, AND TO USE THEIR IMAGINATIONS.

THE SYLLABUS FOR GENERAL BIOLOGY II

GENERAL BIOLOGY II   BIO. 164   SPRING

CREDITS: 4   LECTURE: DAYS: _____   TIME:_____
                          ROOM:____
              LABORATORY:DAY:_____   TIME: _____
                          ROOM:_____

LECTURER: NAME _____ OFFICE _____ PHONE _____
          OFFICE HOURS _____

LABORATORY INSTRUCTOR: NAME _____ OFFICE _____
                       PHONE _____
                       OFFICE HOURS _____

WEEK DATE   SUBJECT   TEXT   LABORATORY

1	INTRODUCTION LEVELS OF ORGANIZATION	INTRODUCTION, MUSEUM TRIP PECHENIK CHAP. 1
2	NUTRITION	#20 PROTISTA PECHENIK CHAP. 2
3	GAS EXCHANGE	#19 MONERA PECHENIK CHAP. 9
4	RESPIRATION	#30 FUNGI
5	EXAM I HOMEOSTASIS	#35 SEED VASCULAR PLANTS
6	OSMOREGULATION	#8 ANIMAL TISSUES, HANDOUT
7	EXCRETION THERMOREGULATION	REVIEW
8	TRANSPORT	MID-TERM
9	EXAM II IMMUNITY	#15 FLOWERING PLANT DEVELOPMENT
10	CHEMICAL INTEGRATION	MUSEUM REPORT DUE, ECOLOGY GAME
11	HORMONES	WATERFALL FIELD TRIP
12	NEURAL INTEGRATION	PIG DISSECTION
13	EXAM III MOTILITY	PIG DISSECTION
14	EVOLUTION - SPECIATION POPULATION	PIG DISSECTION
15	ECOLOGY	LAB. FINAL
16	FINAL	LAB. BOOK DUE

NOTE: THE TOPICS OF REPRODUCTION, DEVELOPMENT AND TAXONOMY ARE COVERED IN THE LABORATORY AND IF TIME PERMITS IN LECTURE.

LECTURE TEXT: 1. TITLE, AUTHOR, EDITION, DATE, PUBLISHER.
REQUIRED

LABORATORY TEXT: 1. EXERCISES IN BIOLOGY, HEIDMANNM, 1984, PWS-

WILLARD GRANT PRESS. REQUIRED
2. A SHORT GUIDE TO WRITING ABOUT BIOLOGY, J.A. PECHENIK, 1987, LITTLE BROWN. REQUIRED
3. DISSECTION OF THE FETAL PIG BY WALKER. REQUIRED

MATERIALS: 1. DRAWING PAPER
2. DISSECTING INSTRUMENTS

GRADE DETERMINATION:
LECTURE: THE LECTURE COMPONENT OF THE COURSE WILL COUNT FOR 3/4 OF THE TOTAL COURSE GRADE. THE LECTURE GRADE WILL BE BASED ON YOUR TWO HIGHEST HOURLY EXAM GRADES PLUS THE CUMULATIVE FINAL. I WILL DROP THE LOWEST OF THE FIRST THREE HOURLY EXAMS AND COUNT ONLY THE HIGHEST TWO. THE FINAL EXAM MUST BE TAKEN. THE FINAL EXAM IS CUMULATIVE: 50% OLD MATERIAL (FROM THE FIRST THREE HOURLY EXAMS) AND 50% NEW MATERIAL (MATERIAL COVERED SINCE THE THIRD EXAM). PART OF LECTURE GRADE WILL BE BASED ON YOUR HOMEWORK ASSIGNMENTS, GROUP WORK, STUDY GROUP ASSIGNMENTS, AND CLASS DISCUSSION.

PART OF THE LECTURE EVALUATION MAY BE THE LOGS.
LOGS: LOGS ARE DUE AT THE BEGINNING OF THE SECOND LECTURE SESSION OF THE WEEK. LATE LOGS WILL NOT BE ACCEPTED. SEE HANDOUT (SEE APPENDIX I, ASSIGNMENT #1).

LABORATORY: THE LABORATORY COMPONENT OF THE COURSE WILL COUNT FOR 1/4 OF THE TOTAL COURSE GRADE. YOU MUST PASS LAB TO PASS THE COURSE. AN "F" IN LAB MEANS AN "F" IN THE COURSE. THE LABORATORY GRADE WILL BE BASED ON TWO LAB EXAMS (MID-TERM AND FINAL), LAB NOTEBOOK, JOURNALS, AND ANY OTHER ASSIGNMENTS MADE BY YOUR INSTRUCTOR. SEE LABORATORY HANDOUTS FOR MORE INFORMATION, CLARIFICATION AND DETAIL.

EXAM POLICY: THERE WILL BE NO MAKE-UP EXAMS AND NO EXTRA CREDIT. CREDIT WILL BE LOST ON ANY QUESTION WHERE SCIENTIFIC WORDS ARE MISSPELLED. SPELLING COUNTS.

TEACHING METHODOLOGIES: I USE A VARIETY OF METHODS IN CLASS; LECTURE, CLASS DISCUSSION, INDIVIDUAL/GROUP WRITING AND DISCUSSION, STUDY GROUPS, LAB PARTNERSHIPS/TEAMS AND SEVERAL TYPES OF WRITING ACTIVITIES. I TRY TO CREATE AN ENVIRONMENT WHICH IS NURTURING, SUPPORTIVE AND NON-COMPETITIVE. I TRY TO ESTABLISH A SENSE OF COMMUNITY THROUGH PEER GROUPS, STUDY GROUPS, LAB PARTNERSHIPS, LAB TEAMS AND THE USE OF THE LOG.

GOALS AND OBJECTIVES: THE GOALS AND OBJECTIVES FOR THIS COURSE ARE NUMEROUS. IN TERMS OF CONTENT THE GOAL IS TO ESTABLISH A COMMON BASE OF BIOLOGICAL KNOWLEDGE (VOCABULARY, CONCEPTS AND THEORIES) FOR THE STUDENTS TO BUILD ON IN LATER COURSES. THIS COURSE GIVES THE SCIENCE MAJORS A SIMILAR BACKGROUND SO THAT OTHER DEPARTMENTAL FACULTY CAN ASSUME A COMMON

LEVEL OF SOCIALIZATION. THERE ARE SEVERAL RECURRENT THEMES THAT RUN THROUGH THE COURSE AND I STRESS AND TIE THEM IN WITH THE LECTURE MATERIAL AT EVERY OPPORTUNITY.

RECURRENT THEMES:
- THE THEORY OF EVOLUTION AND ITS IMPORTANCE TO BIOLOGY
- THE IDEA OF HIERARCHICAL STRUCTURE AND THE INTERCONNECTIONS BETWEEN LEVELS OF COMPLEXITY
- THE IDEA THAT THE WHOLE IS GREATER THAN THE SUM OF ITS PARTS
- THAT BIOLOGY IS A SYNTHETIC SCIENCE
- HISTORICAL PERSPECTIVE OF SCIENCE
- THE RELATIONSHIP BETWEEN STRUCTURE AND FUNCTION
- FEEDBACK MECHANISMS AND CONTROL
- THE EXCITEMENT AND WONDER OF NATURE AND SCIENCE
- THE IMPORTANCE OF THE QUESTIONING APPROACH
- BIOLOGY AS STATE OF THE ART
- THE IMPORTANCE OF SITUATIONAL ETHICS.

I STRUCTURE THE COURSE SO THAT THE STUDENTS NOT ONLY LEARN CONTENT BUT CAN INCREASE THEIR ACADEMIC ABILITIES AND COGNITIVE DEVELOPMENT.

THERE ARE SEVERAL OTHER OBJECTIVES AND GOALS WHICH ARE EMPHASIZED IN THE LABORATORY. THE STUDENTS LEARN HOW TO: DESIGN EXPERIMENTS, FORM AND TEST AN HYPOTHESIS, OBSERVE AND RECORD OBSERVATIONS, IDENTIFY ASSUMPTIONS, MAKE INFERENCES, MANIPULATE DATA AND PRESENT IT IN GRAPHICAL REPRESENTATION, WORK ALONE AND WITH A PARTNER, RELATE BIOLOGY TO OTHER PARTS OF THEIR LIVES, HAVE A BROADER PERSPECTIVE, AND TO USE THEIR IMAGINATIONS.

GENERAL BIOLOGY LABORATORY INFORMATION

THE FOLLOWING IS SOME GENERAL INFORMATION ABOUT THE LABORATORY NOTEBOOK, MATERIALS NEEDED FOR CLASS, EVALUATION CRITERIA, JOURNALS, LAB EVALUATIONS, DRAWINGS, LATE ASSIGNMENTS, AND SOME BASIC VOCABULARY.

LABORATORY NOTEBOOKS:
AT THE END OF THE SEMESTER YOU ARE REQUIRED TO HAND IN YOUR LABORATORY NOTEBOOK. THE NOTEBOOK WILL INCLUDE ALL THE MATERIALS FROM THE LABORATORY: MANUAL EXERCISES, WEEKLY LAB EVALUATIONS, LAB ASSIGNMENTS, LAB JOURNALS, REPORTS, DRAWINGS, WORKSHEETS, EXAMS, ANYTHING YOU DID DURING THE SEMESTER. THE NOTEBOOK WILL BE ORGANIZED WITH ALL THE MATERIAL FROM ONE LAB SESSION TOGETHER. (I SUGGEST THAT YOU USE A THREE RING BINDER). THE LAB NOTEBOOK WILL BE EVALUATED. IF YOU DO NOT HAND IN YOUR NOTEBOOK YOU WILL RECEIVE A GRADE OF INCOMPLETE IN THE COURSE.

MATERIALS FOR CLASS:
WHAT YOU NEED TO BRING TO EACH LABORATORY SESSION: PENCILS, ERASER, THE EXERCISE FOR THE DAY, TEXTBOOK, DRAWING PAPER, ANY

ASSIGNED WORK, AND THE LAB EVALUATION FOR THE LAST WEEKS
LABORATORY.

Evaluation Criteria:
YOUR LAB GRADE WILL BE BASED ON YOUR LAB NOTEBOOK, MID-TERM
GRADE, FINAL GRADE, LAB JOURNALS, AND ALL OTHER ASSIGNED WORK.
YOU ARE RESPONSIBLE FOR READING THE LAB EXERCISE BEFORE COMING TO
CLASS. YOU ARE ALSO RESPONSIBLE FOR READING ANY MATERIAL IN THE
TEXT WHICH COMPLEMENTS THE LAB.

YOU MUST PASS THE LABORATORY PORTION OF THE THIS COURSE TO
PASS THE ENTIRE COURSE AND YOU MUST HAND IN YOUR LABORATORY
NOTEBOOK AND LAB JOURNALS AT THE END OF THE SEMESTER FOR
EVALUATION TO COMPLETE YOUR LAB REQUIRED WORK.

LABORATORY JOURNALS: SEE HANDOUT (SEE APPENDIX I, ASSIGNMENT #2)

LABORATORY EVALUATIONS: SEE HANDOUT (SEE APPENDIX I, ASSIGNMENT
#3)

DRAWINGS: SEE HANDOUT (SEE APPENDIX I, ASSIGNMENT #4)

LATE Assignments:
ANY ASSIGNMENT HANDED IN LATE WILL RECEIVE A LOWER GRADE.
THERE ARE NO MAKE-UP EXAMS.

Basic Vocabulary:
SEE - TO PERCEIVE BY EYE
OBSERVE - TO SEE OR SENSE THROUGH DIRECTED, CAREFUL ANALYTIC
ATTENTION
PROCESS - WHAT YOU DID
PURPOSE - WHY YOU DID IT
ASSUME - TO TAKE FOR GRANTED
INFER - TO ARRIVE AT A CONCLUSION FROM FACTS AND/OR PREMISE
(YOU SEE SMOKE AND INFER A FIRE)
CONCLUDE - A REASONED JUDGMENT (INFERENCE)

Assignments for General Biology I and II

THE PURPOSES OF THE ASSIGNMENTS ARE TO HELP THE STUDENTS WORK
WITH AND LEARN THE CONTENT OF THE COURSE, ENHANCE THEIR ACADEMIC
SKILLS AND INCREASE THEIR COGNITIVE DEVELOPMENT. ALTHOUGH THE
ASSIGNMENTS AND THEIR IMPLEMENTATION ARE PLANNED AHEAD OF TIME
THE ACTUALLY IMPLEMENTATION OF THE ASSIGNMENTS IS NEVER EXACTLY
THE SAME. BECAUSE EACH CLASS IS UNIQUE I REMAIN FLEXIBLE AND
SPONTANEOUS IN CLASS. THE FOLLOWING ASSIGNMENT PLANNING
WORKSHEET WAS USED TO DESIGN ALL OF THE COURSE ASSIGNMENTS (SEE
APPENDIX I, II, III).

Assignment Planning Worksheet

ASSIGNMENT NUMBER _____ TOPIC/LAB. _____

1. PURPOSE:
   TEACHER:
   STUDENT:

2. HOW DOES THE ASSIGNMENT FIT WITH THE OBJECTIVES OF THE COURSE?

3. HOW DOES THE ASSIGNMENT RELATE TO OTHER ASSIGNMENTS?

4. WHAT IS THE AUDIENCE FOR THE ASSIGNMENT?

5. THE ASSIGNMENT.

6. HOW WILL I IMPLEMENT THE ASSIGNMENT IN CLASS?

7. DATE DUE _____ LENGTH (PAGES AND WORDS) _____
   TYPED ___ OR WRITTEN ____

8. WILL THERE BE ANY SUPPORTING MATERIALS OR ROUGH DRAFTS HANDED
   IN WITH THE ASSIGNMENT?_____

9. ASSESSMENT/EVALUATION

   HOW AM I GOING TO ASSESS/EVALUATE THIS ASSIGNMENT?
   CRITERIA:
   SCORING GUIDE:

10. HAVE I PRE-TESTED THE ASSIGNMENT? _____ HOW?_____

11. IS THE HANDOUT READY? _____

    DO I HAVE COPIES OF PREVIOUS EXAMPLES OF STUDENT RESPONSES TO
    THIS ASSIGNMENT TO HAND OUT IN CLASS? _____

AFTER THE ASSIGNMENT HAS BEEN DONE.

12. ARE THERE ANY CHANGES TO BE MADE IN THIS ASSIGNMENT?

13. DO I HAVE NEW EXAMPLES OF STUDENT RESPONSES FOR THIS
    ASSIGNMENT?
                  YES_____   NO _____

CHAPTER 13

THE UNIT APPROACH TO TEACHING LABORATORIES

WHAT IS IT?

WHEN I DESIGN A COURSE WHICH HAS BOTH LECTURE AND LABORATORY I TRY TO COORDINATE THE MATERIAL SO THAT A TOPIC IS COVERED IN LECTURE BEFORE THE STUDENTS NEED TO USE THE INFORMATION IN LABORATORY. SOMETIMES (VERY SELDOM) THIS ACTUALLY OCCURS BUT MOST FREQUENTLY THE LECTURE LAGS FAR BEHIND THE LABORATORY WORK. EVEN A SYLLABUS THAT LOOKS COORDINATED AT THE BEGINNING OF THE SEMESTER TENDS TO FALL OUT OF SINC AS THE SEMESTER PROGRESSES. LECTURE MATERIAL ALWAYS TAKES LONGER TO COVER THAN THE ALOTTED TIME, WHILE THE LABS WHICH ARE COOKBOOK BY NATURE GET SQUASHED INTO A SINGLE 2-3 HOUR TIME SLOT AND THE NEXT WEEK THE LABORATORY CLASS IS ON TO SOMETHING NEW. THE UNIT APPROACH SEEKS TO REMEDY THESE PROBLEMS BY MAKING THE LABORATORY EXPERIENCE SELF-CONTAINED AND THEREFORE NOT TIED DIRECTLY TO THE LECTURE SCHEDULE. DEPENDING ON THE BACKGROUND AND ABILITIES OF THE CLASS THE UNIT CAN BE EXPANDED OR CONTRACTED IN DURATION. ALTHOUGH, IT IS TRUE THAT WITH THE UNIT APPROACH YOU MIGHT NOT COVER AS MANY LABORATORY EXERCISES/EXPERIMENTS IN A SEMESTER OR A YEAR, THE STUDENTS ARE GIVEN THE OPPORTUNITY TO SEE THE WHOLE PICTURE, ACTUALLY HAVE THE TIME TO DESIGN, RUN, EVALUATE AND PRESENT THEIR OWN EXPERIMENTS/EXERCISES, AND THEREFORE ARE IN A BETTER POSITION TO SYNTHESIZE THE WHOLE EXPERIENCE.

LABORATORY MANUALS TODAY ARE ALL QUITE SIMILAR TO EACH OTHER AND ARE VERY MUCH LIKE COOKBOOKS, EACH LAB EXPERIMENT/EXERCISE IS BROKEN DOWN INTO:
1. INTRODUCTION (USUALLY INSUFFICIENT)
2. HOW TO DO THE EXPERIMENT/EXERCISE (STEP BY STEP)
3. 3 OR 4 SHORT ANSWER QUESTIONS ABOUT WHAT THE STUDENTS HAVE DONE.

THE STUDENTS EXPERIENCE EACH LAB AND OFTEN EACH PART OF THE LAB AS A SEPARATE AND AN ENTIRELY SELF-CONTAINED ENTITY. THIS APPROACH TO THE MATERIAL DOES NOT ALLOW THE STUDENTS TO A) SYNTHESIZE THE INFORMATION IN TERMS OF HIERARCHICAL STRUCTURE, B) UNDERSTAND STRUCTURAL AND FUNCTIONAL RELATIONSHIPS, C) CONNECT HOW EACH EXPERIENCE FITS INTO THE SCIENCE, OR D) UNDERSTAND WHAT ACADEMIC SKILLS THEY ARE PRACTICING OR LEARNING.

I HAVE LOOKED FOR BUT HAVE NOT SUCCEED IN FINDING A LABORATORY MANUAL WHICH WOULD HELP THE STUDENTS UNDERSTAND HOW THE MATERIAL/INFORMATION THEY LEARN IN LABORATORY FITS INTO THE WHOLE PICTURE OF BIOLOGY. THEREFORE I WOULD LIKE TO DISCUSS A DIFFERENT WAY OF TEACHING LABORATORIES CALLED THE UNIT APPROACH.

ALTHOUGH AT PRESENT THERE IS NO LABORATORY MANUAL ORGANIZED AROUND THE UNIT APPROACH, ANY MANUAL CAN BE USED SIMPLY BY

PULLING IT APART, REORGANIZING IT INTO UNITS AND PRESENTING THE
UNITS AS AN ORGANIC WHOLE. OF COURSE YOU WILL HAVE TO DESIGN THE
CONNECTING ASSIGNMENTS.

How Does It Work?

A UNIT IS HIERARCHICAL IN STRUCTURE AND CONSISTS OF A GROUP
OF EXPERIMENTS AND/OR EXERCISES THAT ARE FOCUSED ON ONE ORGAN
SYSTEM AND INCLUDES FOR THAT SYSTEM ALL THE LEVELS OF
ORGANIZATION FORM ORGANISMAL TO MOLECULAR. THE UNIT PROGRESSES
FORM THE MORE FAMILIAR/RECOGNIZABLE (THE ORGAN SYSTEM) TO THE
UNFAMILIAR/MORE ABSTRACT/NONE VISIBLE (THE CELLULAR AND
MOLECULAR). THE AMOUNT OF TIME NECESSARY TO COMPLETE A UNIT WILL
VARY ACCORDING TO THE SYSTEM BEING STUDIED AND THE LEVEL OF
PREPARATION OF THE STUDENTS BUT TYPICALLY A UNIT WILL TAKE FROM
FOUR TO FIVE WEEKS TO COMPLETE. EACH UNIT BEGINS WITH THE
DISSECTION OF AN ORGANISM AT THE ORGAN SYSTEM LEVEL, PROGRESSES
TO THE TISSUE LEVEL, THEN TO THE CELLULAR-MOLECULAR LEVEL.
STRUCTURAL-FUNCTIONAL RELATIONSHIPS ARE STRESSED AT EACH LEVEL
BOTH IN DISCUSSION AND IN THE WRITTEN WORK. EACH UNIT CAN BE
STRUCTURED TO SHOW HOW VARIOUS ORGANISMS HAVE SOLVED THE SAME OR
SIMILAR PROBLEMS. THE COMMONALITIES AND DIFFERENCES OF THE
SOLUTIONS CAN BE STRESSED BY DOING COMPARATIVE DISSECTIONS. AT
THE CONCLUSION OF EACH UNIT THE STUDENTS DESIGN AND RUN THEIR OWN
EXPERIMENTS/EXERCISES. USING THEIR OWN EXPERIMENTS/EXERCISES AND
RESULTS THE STUDENTS THEN TIE THE WHOLE UNIT TOGETHER BY ORAL
PRESENTATIONS WHICH ARE FOLLOWED BY CLASS DISCUSSION. THE
EXPERIMENTS/EXERCISES AND RESULTS ARE ALSO PRESENTED TO THE
INSTRUCTOR IN WRITTEN FORM. THE UNIT APPROACH REQUIRES THAT THE
STUDENTS REALLY LEARN TO "SEE" (OBSERVE), RECORD WHAT THEY "SEE",
SYNTHESIS INFORMATION, CARRY OUT DEDUCTIVE AND INDUCTIVE THOUGHT
PROCESSES, MANIPULATED DATA, SOLVE PROBLEMS, WORK WITH A PARTNER
AND EXPRESS WHAT THEY HAVE LEARNED IN VERBAL AND WRITTEN FORM.
BY USING THE UNIT APPROACH STUDENTS SHOULD REACH A BETTER
UNDERSTANDING OF THE INTERRELATIONSHIP OF THE LEVELS OF
HIERARCHICAL STRUCTURE AS WELL AS THE COMPLEXITIES OF THE
STRUCTURAL-FUNCTIONAL RELATIONSHIPS.

UNITS CAN INCORPORATE BOTH PLANT AND ANIMAL MATERIAL FOR THE
SAME SYSTEM AS WELL AS COMPARISONS BETWEEN DIFFERENT PLANTS AND
ANIMALS. SINCE ANY MANUAL CAN BE REASSEMBLED TO FIT THE UNIT
APPROACH EACH INSTRUCTOR CAN USE THE MANUAL OR MATERIALS THEY
LIKE BEST.

METHODS, INSTRUMENTS, TECHNIQUES RECORD   (MIT RECORD)

SINCE STUDENTS WILL BE DESIGNING THEIR OWN EXPERIMENTS OR
EXERCISES OVER THE WHOLE YEAR I STRONGLY RECOMMEND THE KEEPING OF
A METHODS, INSTRUMENT, TECHNIQUES RECORD (MIT RECORD) FOR
REFERENCE. THE MIT RECORD CAN BE KEPT FOR THE DURATION OF THE
UNDERGRADUATE EXPERIENCE AS IT HELPS DEMONSTRATE THAT WHAT
STUDENTS LEARN IN ONE CLASS IS NOT UNIQUE TO THAT CLASS BUT CAN

BE USED LATER. THE FOLLOWING IS THE MIT HANDOUT I GIVE MY CLASS.

MIT RECORD

GENERAL BIOLOGY I AND II ARE THE BEGINNING OF YOUR COLLEGE SCIENTIFIC EDUCATION AND TRAINING WHICH WILL CONTINUE LONG AFTER YOU GRADUATE. DURING THE TIME YOU ARE HERE YOU WILL DO INNUMERABLE EXPERIMENTS/EXERCISES USING MANY DIFFERENT KINDS OF METHODS, INSTRUMENTS, AND TECHNIQUES. YOU CAN'T BE EXPECTED TO REMEMBER ALL OF THESE WITHOUT SOME KIND OF AID - A RECORD. THIS RECORD WILL ALLOW YOU TO HAVE AT YOUR FINGER TIPS FOR EASY REFERENCE A LIST OF METHODS, INSTRUMENTS, TECHNIQUES AND THEIR VARIOUS USES. WHEN YOU APPROACH A RESEARCH PROBLEM AND NEED TO DESIGN AN EXPERIMENT YOU WILL HAVE THIS RECORD TO REFER TO AND WORK WITH. SOME OF THE METHODS, INSTRUMENTS AND TECHNIQUES WILL BE USED MORE THAN ONCE, THESE EXPERIENCES SHOULD BE KEPT ON THE SAME PAGE OR ON CONSECUTIVE PAGES. THE MIT RECORD WILL BE HANDED IN TWICE A SEMESTER (AT THE TIME OF THE MID-TERM EXAM AND AT THE TIME OF THE FINAL EXAM). I SUGGEST THAT YOU USE A LOOSE LEAF, THREE RING BINDER SO THAT YOU CAN ADD AND REARRANGE PAGES AS YOU PROCEED.

THE PROBLEM IS HOW TO SET UP SUCH A RECORD. WHAT TYPES OF INFORMATION ARE WE INTERESTED IN RECORDING? HOW CAN THE INFORMATION BE QUICKLY RECORDED AND STILL BE EASILY RETRIEVED?
THE FOLLOWING ARE SOME QUESTIONS THAT NEED TO BE ANSWERED:
1. IN WHAT CLASS DID YOU DO THE EXPERIMENT/EXERCISE (THIS ALLOWS YOU TO REFER BACK TO THE CLASS NOTEBOOK)?

2. WHAT WAS THE NAME OF THE EXPERIMENT/EXERCISE?

3. WHAT WAS THE METHOD, INSTRUMENT, OR TECHNIQUE (MIT) USED?

4. WHAT WAS THE MIT USED FOR (IN OTHER WORDS WHAT INFORMATION DID YOU OBTAIN)?

5. WHAT CAN THE MIT BE USED FOR IN A MORE GENERAL SENSE?

6. HOW DID THE MIT WORK?

NOTE: THE TERM TECHNIQUES INCLUDES WAYS OF HANDLING DATA.

MIT FORMAT

REFERENCE - INCLUDES DATE OF USE, COURSE, NAME OF EXPERIMENT OR EXERCISE

WHAT MIT - INCLUDES THE NAME OF THE MIT, HOW IT WORKS

PURPOSE - INCLUDES WHAT KIND OF INFORMATION CAN BE OBTAINED, (THIS SHOULD BE SPECIFIC FOR THE EXPERIMENT OR

EXERCISE AND THEN GENERALIZED)

EXAMPLES:

REFERENCE	WHAT MIT	PURPOSE
85, GBI ENZYME EXP.	SPECTROMETER 20/21 SHINES LIGHT THROUGH LIQUID TO MEASURE THE AMOUNT OF LIGHT TRANSMITTED OR ABSORBED	USED TO DETERMINE PROTEIN CONCENTRATION
		CAN BE USED TO MEASURE THE CONCENTRATION OR AMOUNT OF ANY MATERIAL IN
	NEEDS A PIGMENT TO WORK	SOLUTION
85, GBI ENZYME EXP.	GRAPH-BEST FIT LINE PLOT DATA AND PLACE LINE SO THAT + AND - POINTS ARE EVENLY DISTRIBUTED AROUND THE LINE	USED FOR CONTINUOUS DATA HELPS TO VISUALIZ THE RATE OF THE PROCESS WHEN YOU HAVE LITTLE DATA AND A LOT OF VARIATION
85, GBI PHOTOSYNTHESIS EXP.	SPECTROMETER 20/21 SEE ABOVE (PLACE WITH OTHER RECORD)	USED TO IDENTIFY WHICH WAVELENGTH ABSORBED BY EACH PIGMENT
		IF YOU KNOW THE WAVELENGTH YOU CAN IDENTIFY THE PIGMENT.

Sample Units for General Biology I and II

THESE SAMPLE UNITS COULD BE PUT TOGETHER USING THE EXISTING LABORATORY MANUAL AND ASSIGNMENTS. HOWEVER, THERE WOULD HAVE TO BE A FAIR AMOUNT OF DISCUSSION WITH THE LABORATORY PARTNERS/TEAMS WHEN IT CAME TIME FOR THEM TO DESIGN AN EXPERIMENT.

Unit I: OBSERVATION/RECORDING

THIS UNIT INTRODUCES THE STUDENTS TO THE BASIC LABORATORY TOOL - THE MICROSCOPE, THE SKILLS OR ABILITIES OF OBSERVATION AND RECORDING.

OBSERVATION AT THE MACRO LEVEL.
1. A FIELD TRIP TO SEE AN ECOSYSTEM, TO DISCUSS ITS STRUCTURE AND FUNCTION AND EVOLUTION.
2. THE ECOLOGY GAME.

OBSERVATION AT THE ORGANISMAL AND THE MICRO LEVEL.
1. THE STUDENTS USE THE DISSECTING AND LIGHT MICROSCOPES AND LEARN HOW TO RECORD WHAT THEY SEE (DRAWING). THIS INCLUDES HOW TO USE AN OCULAR MICROMETER.

2. They are given demonstrations of how the Transmission and Scanning Electron Microsacopes work and learn to look at micrographs and determine actual size.

Unit II: ENERGY/DIGESTION/PHOTOSYNTHESIS

ANIMAL
1. Organ system - dissection of the fetal pig and comparison organisms.
2. Dissecting and light microscope observation of hand cut samples of the various parts of the organ system.
3. Light microscope observation of prepared tissue slides.
4. Chemical structure of carbohydrates.
5. Chemical structure and function of proteins and enzymes.
6. Students design, run and present experiments or exercises.

PLANT
1. Organ system - leaf and comparison organisms.
2. Dissecting and light microscope observation of hand cut samples of the various parts of a leaf.
3. Light microscope observation of prepared organ and tissue slides.
4. Exercise on whole organ photosynthesis.
5. Exercise on pigment extraction and absorption spectrum.
6. Students design, run and present experiments or exercises.

Unit III: CIRCULATION/TRANSPORT

ANIMAL
1. Organ system - circulation and excretion - dissection of the fetal pig and comparison organisms.
2. Dissecting and light microscope observation of hand cut samples of the various parts of the organ systems.
3. Light microscope observation of prepared tissue slides.
4. Blood typing and hemolysis.
5. Exercises on diffusion and osmosis.
6. Students design, run and present experiments or exercises.

PLANT
1. Organ system - leaf, root and stem and comparison organisms.
2. Dissecting and light microscope observation of hand cut samples of the various parts of a leaf, root, stem.
3. Light microscope observation of prepared organ and tissue slides.

4. EXPERIMENT ON WATER PRESSURE AND TUGOR.
5. STUDENTS DESIGN, RUN AND PRESENT EXPERIMENTS OR EXERCISES.

UNIT IV: REPRODUCTION/GENETICS

ANIMAL
1. ORGAN SYSTEM - REPRODUCTION - DISSECTION OF THE FETAL PIG AND COMPARISON ORGANISMS.
2. DISSECTING AND LIGHT MICROSCOPE OBSERVATION OF HAND CUT SAMPLE OF THE VARIOUS PARTS OF THE ORGAN SYSTEM.
3. LIGHT MICROSCOPE OBSERVATION OF PREPARED TISSUE SLIDES.
4. EXERCISE ON STAINING PROCEDURE, SLIDES OF MITOSIS AND MEIOSIS.
5. EXERCISES ON MENDELIAN GENETICS.
6. EXERCISES ON HARDY-WEINBERG, POPULATIONAL GENETICS.
7. STUDENTS DESIGN, RUN AND PRESENT EXPERIMENTS OR EXERCISES.

PLANT
1. ORGAN SYSTEM - DISSECTION OF THE FLOWER AND COMPARISON ORGANISMS.
2. DISSECTING AND LIGHT MICROSCOPE OBSERVATION OF HAND CUT SAMPLES OF THE VARIOUS PARTS OF A LEAF.
3. LIGHT MICROSCOPE OBSERVATION OF PREPARED ORGAN AND TISSUE SLIDES.
4. EXERCISE ON SEED GERMINATION.
5. EXERCISE ON COMPARATIVE LIFE CYCLES.
6. STUDENTS DESIGN, RUN AND PRESENT EXPERIMENTS OR EXERCISES.

UNIT V: RESPIRATION/WHOLE ORGANISM/CELLULAR

ANIMAL
1. ORGAN SYSTEM - DISSECTION OF THE FETAL PIG AND COMPARISON ORGANISMS.
2. DISSECTING AND LIGHT MICROSCOPE OBSERVATION OF HAND CUT SAMPLES OF THE VARIOUS PARTS OF THE ORGAN SYSTEM.
3. LIGHT MICROSCOPE OBSERVATION OF PREPARED TISSUE SLIDES.
4. EXERCISE ON WHOLE ORGAN RESPIRATION.
5. STUDENTS DESIGN, RUN AND PRESENT EXPERIMENTS OR EXERCISES.

PLANT
1. ORGAN SYSTEM - LEAF AND COMPARISON ORGANISMS.
2. DISSECTING AND LIGHT MICROSCOPE OBSERVATION OF HAND CUT SAMPLES OF THE VARIOUS PARTS OF A LEAF.
3. LIGHT MICROSCOPE OBSERVATION OF PREPARED ORGAN AND TISSUE SLIDES.
4. EXERCISE ON CELLULAR RESPIRATION.

5. STUDENTS DESIGN, RUN AND PRESENT EXPERIMENTS OR EXERCISES.

REFERENCES

Belenky, Mary Field et al. 1986. Women's Ways of Knowing: The Development of Self, Voice and Mind. Basic Books Inc, Publisher, New York. pp. 256.

Braid, Bernice. 1986. Personal Communication. William Paterson College, Seminar.

Chickering, Arthur W. and Associates. 1981. The Modern American College. Jossey-Bass Publishers, San Francisco. 4th ed. pp. 810.

Eisley, Loren. 1975. All The Strange Hours. Charles Schribner's Sons, New York. pp. 266.

Flower, Linda. 1982. Problem-Solving Strategies for Writing. Harcourt Brace Jovanoick Inc., New York. pp. 210.

Furman, Barbara Schneider and Anthony F. Grasha. 1983. A Practical Handbook for College Teachers. Little Brown and Company, Canada. pp. 315.

Hanks, Sharon R. 1985. Writing as Problem Solving. Chapter 5 in On Writing Well: A Faculty Guidebook for Improving Student Writing in all Disciplines by Kloss et al. William Paterson College, New Jersey. pp. 23-29.

Hayes, John R. 1981. The Complete Problem Solver. The Franklin Institute Press, Philadelphia. pp. 255.

Judy, Stephen N. and Susan J. Judy. 1981. An Introduction to the Teaching of Writing. John Wiley and Sons, New York. pp. 193.

Kloss, Robert J. et al. 1985. On Writing Well: A Faculty Guidebook for Improving Student Writing in all Disciplines. William Paterson College, New Jersey. pp. 58.

Kolb, David A. 1981. Learning Styles and Disciplinary Differences. Chapter 10 in The Modern American College by Chickering. Jossey-Bass Publishers, San Francisco. 4th ed. pp. 232-255.

Magolda, Marcia B. 1985. Personal Communication. William Paterson College, Seminar.

McKeachie, Wilbert J. 1986. Teaching Tips: A Guidebook for the Beginning Teacher. D.C. Heath and Company. Lexington. 8th ed. pp. 53.

Meyers, Chet. 1986. Teaching Students to Think Critically. Jossey-Bass Publishers, San Francisco. pp. 131.

NEW JERSEY BASIC SKILLS COUNCIL. MARCH 19, 1986. THINKING SKILLS AN OVERVIEW. DEPARTMENT OF HIGHER EDUCATION, TRENTON.

PARKER, ROBERT P. AND VERA GOODKIN. 1987. THE CONSEQUENCES OF WRITING: ENHANCING LEARNING IN THE DISCIPLINES. BOYTON/COOK PUBLISHERS, INC. UPPER MONTCLAIR, NEW JERSEY. PP. 183.

PECHENIK, JAN A. 1987. A SHORT GUIDE TO WRITING ABOUT BIOLOGY. LITTLE BROWN AND COMPANY, BOSTON. PP. 194.

PERRY, WILLIAM. 1981. COGNITIVE AND ETHICAL GROWTH: THE MAKING OF MEANING. CHAPTER 3 IN THE MODERN AMERICAN COLLEGE BY CHICKERING. JOSSEY-BASS PUBLISHERS, SAN FRANCISCO. 4TH ED. PP. 76-116.

TCHUDI, STEPHEN N. 1986. TEACHING WRITING IN CONTENT AREAS. NATIONAL EDUCATION ASSOCIATION OF THE UNITED STATES. PP. 128.

WHITE, EDWARD M. 1986. TEACHING AND ASSESSING WRITING. JOSSEY-BASS PUBLISHERS, SAN FRANCISCO. PP. 304.

WIDICK, CAROLE AND DEBORAH SIMPSON. 1978. DEVELOPMENTAL CONCEPTS IN COLLEGE INSTRUCTION. IN ENCOURAGING DEVELOPMENT IN COLLEGE STUDENTS EDITED BY C. PARKER. UNIVERSITY OF MINNEAPOLIS UNIVERSITY PRESS, MINNEAPOLIS. PP. 27-59.

WILLIAM, JOE. 1986. PERSONAL COMMUNICATION. WILLIAM PATERSON COLLEGE, SEMINAR.

APPENDIX I

ASSIGNMENTS FOR GENERAL BIOLOGY I AND II SYLLABI (#1-4)

ASSIGNMENT # 1

GENERAL BIOLOGY I AND II

LECTURE   TOPIC   LOG

THE ASSIGNMENT:

A LOG IS A 5 MINUTE WRITING EXERCISE DUE AT THE BEGINNING OF THE SECOND LECTURE SESSION OF EACH WEEK. LOGS ARE VOLUNTARY.

WHAT IS THE PURPOSE OF THE LOG EXPERIENCE? THE PURPOSE OF THE LOG FOR ME IS SO THAT I CAN GET TO KNOW YOU BETTER AND PERHAPS ANSWER ANY QUESTIONS YOU MAY HAVE DURING THE SEMESTER. FOR THE CLASS THE LOGS CAN FOSTER A FREER EXCHANGE OF IDEAS AS YOU (THE STUDENTS) MAY FEEL THAT IT IS EASIER TO ASK QUESTIONS AND DISCUSS IDEAS. FOR YOU, PERSONALLY, THIS IS A CHANCE TO REFLECT ON AND EXPRESS WHAT IS GOING ON IN YOUR LIFE.

FIVE MINUTES OF WRITING IS APPROXIMATELY 1/2 TO 1 PAGE OF WRITING (@ 150-250 WORDS). LOGS MAY BE TYPED OR WRITTEN BY HAND. YOUR LOGS ARE NOT GRADED BUT I KEEP A RECORD OF HOW MANY YOU HAVE HANDED IN DURING THE SEMESTER. LOGS ARE ENTIRELY CONFIDENTIAL BETWEEN YOU AND ME. I RETURN YOUR LOG TO YOU AT THE BEGINNING OF THE NEXT LECTURE SESSION.

YOU MAY WRITE ON ANY SUBJECT. THE FOLLOWING ARE A FEW EXAMPLES OF WHAT STUDENTS HAVE WRITTEN ABOUT IN PAST CLASSES: QUESTIONS NOT ASKED IN CLASS, QUESTIONS ABOUT THE COURSE, PRACTICE ESSAY ANSWERS FOR EXAMS, LECTURE SUMMARIES, QUESTIONS ABOUT SCHOOL IN GENERAL, THOUGHTS ABOUT SCHOOL, HOME, JOBS, SPORTS, HOBBIES, FOOD, POEMS AND SONGS WRITTEN BY STUDENTS, THINGS YOU THINK ABOUT IN GENERAL.

IN OTHER WORDS LOGS CAN BE ANYTHING AT ALL.

HOW DO I USE THE LOGS IN TERMS OF COURSE EVALUATION? I USE THE LOGS AS A SWING GRADE, THAT IS IF YOU ARE ON THE BORDER BETWEEN A C+ (79%) AND AN B- (80%) AND YOU HAVE HANDED IN 2/3 OF THE LOGS DURING THE SEMESTER (E.G. 10 OUT OF 15) YOU WILL RECEIVE THE B- BUT IF YOU HAVEN'T TURNED IN THE LOGS YOU GET THE C+.

IF YOU HAVE ANY QUESTIONS ABOUT THE LOGS PLEASE ASK ME.

THE IMPLEMENTATION:

I HAND OUT THE LOG ASSIGNMENT DURING THE FIRST LECTURE SESSION AND DISCUSS IT WITH THE CLASS. I ALSO ASK IF THEY HAVE ANY QUESTIONS AT THE SECOND LECTURE SESSION WHEN THE FIRST LOGS CAN BE HANDED IN.

ASSIGNMENT # 2

GENERAL BIOLOGY I AND II

LABORATORY EXERCISE   LABORATORY JOURNAL

THE ASSIGNMENT:

ONE OF THE LABORATORY REQUIREMENTS THIS SEMESTER IS A JOURNAL. THE PURPOSE OF THE JOURNAL IS TO ENCOURAGE YOU TO EXPERIENCE BIOLOGY OUTSIDE OF THE CLASSROOM, TO REFLECT ON THE EXPERIENCE AND TO ASSESS ITS CONTRIBUTION TO YOUR SCIENTIFIC BACKGROUND OR FRAMEWORK.

YOU WILL EXPERIENCE BIOLOGY DURING THE SEMESTER IN SEVERAL DIFFERENT WAYS (AREAS). YOU MUST DO AT LEAST ONE JOURNAL ENTRY FOR EACH OF THE EIGHT AREAS. THERE WILL BE TEN JOURNAL ENTRIES HANDED IN DURING THE SEMESTER. IF YOU HAVE ANY OTHER IDEAS FOR JOURNAL ENTRIES LET ME KNOW SO THAT I CAN SUGGEST THEM AS ALTERNATIVE AREAS TO THE CLASS.

JOURNAL ENTRIES WILL BE ONE TO TWO PAGES TYPED DOUBLE SPACED OR LEGIBLY WRITTEN (@ 250-500 WORDS).

AREAS

1. WATCH A TV SHOW OR SEE A MOVIE CONCERNED WITH BIOLOGY.
   (E.G. CHANNEL 13 SCIENCE SERIES, CABLE TV SCIENCE SERIES, NEWTON'S APPLE, NATIONAL GEOGRAPHIC SERIES, THE BRAIN SERIES, NATURE SERIES)

2. READ AN ARTICLE IN A POPULAR MAGAZINE ABOUT SOME ASPECT OF BIOLOGY.
   (E.G. TIME, NEWSWEEK, FORTUNE, PEOPLE)

3. READ AN ARTICLE IN A GENERAL SCIENCE JOURNAL ABOUT SOME ASPECT OF BIOLOGY.
   (E.G. SMITHSONIAN, NATURALIST, NATIONAL GEOGRAPHIC, SCIENCE'86, DISCOVERY)

4. READ AN ARTICLE IN A SCIENTIFIC JOURNAL ABOUT SOME ASPECT OF BIOLOGY.
   (E.G. AMERICAN SCIENTIST, SCIENTIFIC AMERICAN, NATURE, SCIENCE, ANY PROFESSIONAL SCIENCE JOURNAL)

5. READ AN ARTICLE IN A DAILY OR WEEKLY NEWSPAPER ABOUT SOME ASPECT OF BIOLOGY.
   (E.G. NY TIMES TUESDAY SCIENCE SECTION, LOCAL NEWSPAPER, NATIONAL NEWSPAPER)

6. INTERVIEW (HAVE A CONVERSATION WITH) SOMEONE WHOSE JOB IS RELATED TO BIOLOGY.
   (E.G. FAMILY DOCTOR, DENTIST, VETERANARIAN, DRUGGIST, PARK

RANGER, TEACHER, NURSE, ZOO ATTENDANT)

7. VISIT ONE OF THE FOLLOWING.
(E.G. MUSEUM OF NATURAL HISTORY, ZOO, GAME RESERVE, BIRD SANCTUARY, ANIMAL SHELTER, BOTANICAL GARDEN, WHALE WATCH, AQUARIUM) ATTACH AN ADMISSION TICKET TO YOUR JOURNAL ENTRY IF POSSIBLE.

8. ATTEND A LECTURE OR SEMINAR ON SOME ASPECT OF BIOLOGY.
(E.G. NY ACADEMY OF SCIENCE, NJ ACADEMY OF SCIENCE)

FOR EACH ENTRY YOU MUST INCLUDE REFERENCE INFORMATION AT THE BEGINNING OF THE ENTRY.

EXAMPLES

TV PROGRAM
CHANNEL _____ DATE _____ TIME _____
NAME OF THE PROGRAM _____

POPULAR MAGAZINE ARTICLE
NAME OF THE MAGAZINE _____ DATE _____
PAGES ____ NAME OF THE ARTICLE _____
AUTHOR _____

INTERVIEW
NAME OF PERSON INTERVIEWED _____ DATE _____
TIME _____ OCCUPATION _____

THE FOLLOWING ARE SOME QUESTIONS YOU SHOULD THINK ABOUT BEFORE YOU START A JOURNAL EXPERIENCE, REFLECT ON AFTER YOU HAVE HAD THE EXPERIENCE, AND INCORPORATE INTO YOU ENTRY.

1- WHAT WAS IT ABOUT?

2- DID I LEARN ANYTHING NEW? DO I NOW SEE SOMETHING IN A DIFFERENT WAY OR FROM A DIFFERENT PERSPECTIVE?

3- DID THE EXPERIENCE RAISE ANY ISSUES FOR ME TO THINK ABOUT? DO I HAVE ANY UNANSWERED QUESTIONS ABOUT THE MATERIAL?

4- WHAT WAS I FEELING WHILE I WAS WATCHING, READING, LISTENING, TALKING OR DOING ?

5- WHAT WAS I FEELING AFTER THE EXPERIENCE WAS OVER?

6- HOW DOES THIS EXPERIENCE FIT INTO MY IDEAS ABOUT BIOLOGY?

ASSESSMENT:

THE JOURNAL WILL BE COLLECTED FOUR TIMES DURING THE SEMESTER. THE COLLECTION DATES FOR THIS SEMESTER ARE: _____, _____,

_____, _____. ALL THE JOURNAL ENTRIES ARE TO BE HANDED IN WITH YOUR LABORATORY NOTEBOOK AT THE END OF THE SEMESTER. JOURNAL ENTRIES ARE ASSESSED ACCORDING TO THE FOLLOWING SCORING GUIDE:

- 3- DID THE ASSIGNED TASK, ENTRY SHOWS A SERIOUS EFFORT
- 2- DID THE TASK, ENTRY DONE HASTILY, BRIEF, SHOWS LITTLE THOUGHT
- 1- CURSORY WORK
- 0- WORK NOT DONE

TO HELP YOU KEEP TRACK OF YOUR JOURNAL ENTRIES THE FOLLOWING RECORD MUST BE HANDED IN AT THE END OF THE SEMESTER WITH YOUR ENTRIES.

JOURNAL RECORD   SEMESTER _____   NAME _____

ENTRY	AREA
1	_____
2	_____
3	_____
4	_____
5	_____
6	_____
7	_____
8	_____
9	_____
10	_____

THE IMPLEMENTATION:

THE JOURNAL HANDOUT IS DISTRIBUTED DURING THE FIRST LAB SESSION AND DISCUSSED. BEFORE THE FIRST JOURNAL ENTRY IS DUE THERE IS ANOTHER CLASS DISCUSSION.

AT THE END OF THE SEMESTER THE JOURNAL WORK WILL BE TRANSLATED INTO AN OVERALL A-F GRADE AS PART OF THE LABORATORY GRADE.

ASSIGNMENT # 3

GENERAL BIOLOGY I AND II

LABORATORY EXERCISE   LABORATORY EVALUATIONS

THE ASSIGNMENT:

THIS SEMESTER YOU WILL BE WRITING A LABORATORY EVALUATION FOR EACH LABORATORY. THE PURPOSE OF THE EVALUATION IS TO GIVE YOU PRACTICE SUMMARIZING INFORMATION, IDENTIFYING THE MAIN PURPOSE OF THE LABORATORY, AND EVALUATING YOUR EXPERIENCE. THESE EVALUATIONS GIVE ME INSTANT FEEDBACK ON THE YOUR LABORATORY EXPERIENCE SO THAT I SEE WHAT PROBLEMS YOU ARE HAVING AND I CAN HELP YOU UNDERSTAND THE MATERIAL BETTER. YOUR EVALUATIONS ALSO HELP ME TO IMPROVE THE LABORATORY EXERCISES/EXPERIMENTS.

LABORATORY EVALUATIONS ARE SHORT, NO MORE THAN 1/2 TO ONE PAGE IN LENGTH (150-250 WORDS). THEY MAYBE TYPED DOUBLE SPACED OR LEGIBLY WRITTEN. THEY ARE COMMENTED ON, ASSESSED AND HANDED BACK THE FOLLOWING WEEK. THE EVALUATIONS ARE TO BE PLACED IN YOUR LABORATORY NOTEBOOK WITH THE CORRESPONDING EXERCISE/EXPERIMENT. THEY CONSTITUTE PART OF THE LAB NOTEBOOK GRADE.

EACH EVALUATION MUST INCLUDE THE FOLLOWING:

1- A SUMMARY IN ONE SENTENCE OF THE PURPOSE OF THE LABORATORY IN TERMS OF THE BIOLOGICAL IDEAS, CONCEPTS OR THEORIES COVERED IN THE LAB SESSION. THIS IS WHY YOU DID THE LABORATORY NOT HOW YOU DID IT (METHODS) OR WHAT YOU USED (MATERIALS).

2- AN OVERALL EVALUATION OF THE LABORATORY (DID YOU FIND THAT IT MET THE OBJECTIVES STATED IN THE LAB MANUAL).

3- SPECIFIC PROBLEMS YOU HAD WITH THE MATERIAL.

4- SPECIFIC SUGGESTIONS FOR IMPROVING THE LABORATORY.

5- A LISTING OF ME WHICH PAGES AND ILLUSTRATIONS OF YOUR TEXT WERE USEFUL IN UNDERSTANDING THE EXERCISES/EXPERIMENTS.

ASSESSMENT:

EVALUATIONS ARE DUE AT THE BEGINNING OF NEXT LABORATORY SESSION.

LABORATORY EVALUATIONS WILL BE ASSESSED ACCORDING TO THE FOLLOWING SCORING GUIDE:

+ = DID THE ASSIGNED TASK, SHOWS A SERIOUS EFFORT

\- = DID THE TASK, DONE HASTILY, BRIEFLY, SHOWS LITTLE THOUGHT
0 = WORK NOT DONE

THE IMPLEMENTATION:

THIS ASSIGNMENT IS HANDED OUT AND DISCUSSED AT THE FIRST LAB SESSION. BEFORE THE EVALUATION IS HANDED IN AT THE NEXT LAB SESSION EACH STUDENT WILL READ THEIR LAB PARTNER'S AND TABLE PARTNERS' EVALUATIONS TO SEE THEIR DIFFERENT POINTS OF VIEW.

AS I READ THE EVALUATIONS I INDICATE WHERE I HAVE QUESTIONS, WHAT ARE THE GOOD POINTS, AND WHICH AREAS NEED WORK.

ASSIGNMENT # 4

GENERAL BIOLOGY I AND II

LABORATORY EXERCISE   LABORATORY DRAWINGS

THE ASSIGNMENT:

WHEN YOU DO BIOLOGICAL DRAWINGS YOU ARE ACTUALLY DOING
SEVERAL THINGS; YOU ARE LEARNING ABOUT BIOLOGY, INCREASING YOUR
OBSERVATIONAL SKILLS, INCREASING YOUR ABILITY TO RECORD YOUR
OBSERVATIONS BOTH IN GRAPHIC AND IN WRITTEN FORM, AND YOU ARE
PUTTING THE INFORMATION INTO LONG TERM MEMORY. YOUR COMPLETED
DRAWINGS AND THE INFORMATION INCLUDED WITH THEM ARE USEFUL STUDY
MATERIALS FOR THE LABORATORY EXAMS. THE MORE COMPLETE YOUR
DRAWINGS ARE THE MORE USEFUL THEY WILL BE AS STUDY GUIDES. A
BRIEF SKETCH IS NOT A DRAWING.

ALL LABORATORY DRAWINGS MUST BE DONE IN PENCIL AND MUST
FOLLOW THE FOLLOWING FORMAT:

ON THE FRONT OF THE PAGE OF EACH DRAWING:

TITLE:_____   YOUR NAME _____

DRAWING SIZE IN MM   _____
ACTUAL SIZE IN UM   _____
TOTAL MAGNIFICATION   _____
MAGNIFICATION FACTOR   _____

(THE DRAWING ITSELF SHOULD BE LARGE, COVERING ABOUT 1/2 TO
   3/4 OF THE PAPER, AND MUST BE FULLY LABELED WITH THE LABELS
   PRINTED IN PENCIL)

SCALE = _____   DATE _____

ON THE BACK OF EACH DRAWING:

DESCRIBE: WRITE A FULL, DETAILED DESCRIPTION OF WHAT YOU HAVE
          OBSERVED. THIS SHOULD INCLUDE SIZE, SHAPE, TEXTURE,
          AND COLOR. THIS IS A WRITTEN RECORD OF WHAT YOU
          OBSERVED IN SENTENCE FORM. THIS IS NOT A LIST.

ASSOCIATE: WHAT DOES THE OBJECT YOU OBSERVED REMIND YOU OF?
          BY ASSOCIATING WHAT YOU OBSERVE WITH SOMETHING
          ELSE IN YOUR EXPERIENCE YOU WILL REMEMBER IT
          LONGER. E.G. A CHEEK CELL REMINDS ME OF A FRIED
          EGG, WITH THE NUCLEUS AS THE YOLK.

COMPARE: FOR EACH DRAWING YOU DO I WILL GIVE YOU ANOTHER
          OBJECT TO OBSERVE FOR COMPARISON. YOUR COMPARISON
          SHOULD TELL ME HOW THE TWO OBJECTS ARE SIMILAR AND

HOW THEY ARE DIFFERENT. AGAIN THIS IS NOT A LISTING
EXERCISE BUT IS DONE IN COMPLETE SENTENCES.

ASSESSMENT:

DRAWINGS ARE GRADED ON COMPLETENESS OF INFORMATION AND
CLARITY NOT ON ARTISTIC ABILITIES. DRAWINGS ARE DUE AT THE
BEGINNING OF THE NEXT LABORATORY SESSION.

THE IMPLEMENTATION:

THE DRAWING INSTRUCTIONS ARE HANDED OUT AT THE FIRST
LABORATORY SESSION AND DISCUSSED WITH THE STUDENTS. THERE IS
ADDITIONAL DISCUSSION WHEN THE FIRST DRAWING ASSIGNMENT IS MADE.

APPENDIX II

WRITING ASSIGNMENTS FOR GENERAL BIOLOGY LABORATORIES (#5-30)

ASSIGNMENT # 5

GENERAL BIOLOGY I

LABORATORY EXERCISE   LABORATORY INTRODUCTION

THE ASSIGNMENT:

THIS IS THE FIRST LABORATORY EXERCISE OF THE SEMESTER.

1- EXPLAIN CLUSTERING USING A SPECIFIC EXAMPLE.

2- DISTRIBUTE PAPER AND HAVE THE STUDENTS INDIVIDUALLY CLUSTER ON THE TERM BIOLOGY LAB. THEN HAVE LAB PARTNERS EXCHANGE CLUSTERS SO THEY CAN SEE HOW ANOTHER PERSON VIEWS THE LABORATORY.

3- EXPLAIN FREE WRITING.

4- DISTRIBUTE PAPER AND HAVE THE STUDENTS FREE WRITE ON BIOLOGY LAB FOR 5 MIN.

5- DIVIDED THE CLASS INTO LAB TEAMS OF FOUR AND HAVE EACH MEMBER OF THE TEAM READ THEIR FREE WRITING ALOUD TO THE TEAM. AFTER GROUP DISCUSSION EACH TEAM CHOOSES ONE WRITING TO PRESENT TO THE CLASS.

6- LAB TEAM MEMBERS EXCHANGE NAMES AND PHONE NUMBERS WITH EACH OTHER AND ONE STUDENT, ACTING AS SECRETARY, MAKES A NAME/PHONE LIST FOR ME.

7- EACH TEAM READS THEIR CHOOSEN FREE WRITING ALOUD TO THE CLASS AND THE CLASS DISCUSSES THE FREE WRITING SAMPLES. THIS DISCUSSION LEADS INTO WHAT LAB IS GOING TO BE ABOUT THIS SEMESTER AND THE RULES OF THE ROAD. I HAND OUT INFORMATION ON THE LAB FORMAT, EVALUATIONS, JOURNALS, AND DRAWINGS.

THE IMPLEMENTATION:

ALL OF THE ASSIGNMENT IS DONE IN CLASS. AT THE END OF THE CLASS I COLLECT THE CLUSTERING AND FREE WRITING EXERCISES. I READ AND MAKE SUPPORTIVE COMMENTS ON THESE PAPERS AND RETURN THEM AT THE NEXT LAB SESSION.

ASSIGNMENT # 6

GENERAL BIOLOGY I

LABORATORY EXERCISE #8   STRUCTURE OF CELLS

THE ASSIGNMENT:

THE PURPOSE OF THIS ASSIGNMENT IS TO:

1- FAMILIARIZE YOU WITH THE STRUCTURE AND FUNCTIONING OF THE LIGHT MICROSCOPE

2- INCREASE YOUR OBSERVATIONAL SKILLS

3- GIVE YOU PRACTICE IN RECORDING WHAT YOU OBSERVED BY DRAWING AND DESCRIBING IT.

DRAW YOUR CHEEK CELL USING THE FORMAT GIVEN OUT LAST WEEK. SINCE YOU HAVE NOT YET LEARNED HOW TO MEASURE THE ACTUAL SIZE OF THE CELL, MAKE A SCALE, OR TO CALCULATE THE MAGNIFICATION FACTOR THESE WILL NOT BE REQUIRED FOR THIS DRAWING. HOWEVER BE SURE TO INCLUDE: TITLE, LABELS, TOTAL MAGNIFICATION, DRAWING SIZE, DATE, DESCRIPTION, ASSOCIATION AND A COMPARISON. DETAILED DESCRIPTIONS, ASSOCIATIONS, AND COMPARISONS ARE ALWAYS WRITTEN IN SENTENCE FORM. COMPARE YOUR CHEEK CELL WITH THE PLANT CELL OBSERVED IN LAB TODAY.

ASSESSMENT:

THIS ASSIGNMENT IS DUE AT THE BEGINNING OF NEXT LABORATORY SESSION.

THE ASSIGNMENT IS WORTH TEN POINTS. LATE ASSIGNMENTS RECEIVE LOWER GRADES.

THE ASSIGNMENT WILL BE ASSESSED ON THE COMPLETENESS OF THE INFORMATION PRESENTED AND ON THE CLARITY AND ACCURACY OF THE DRAWING (THIS DOES NOT MEAN ARTISTIC ABILITY).

THE IMPLEMENTATION:

THIS ASSIGNMENT IS HANDED OUT AT THE BEGINNING OF THE LAB SESSION AND DISCUSSED. THE DRAWING IS COLLECTED AT THE BEGINNING OF THE NEXT LAB SESSION.

ASSIGNMENT # 7

GENERAL BIOLOGY I

LABORATORY EXERCISE   MICROSCOPE LABORATORY

THE ASSIGNMENT:

THE PURPOSE OF THIS ASSIGNMENT IS TO GIVE YOU PRACTICE IN:

1- TRANSLATING WORD PROBLEMS INTO MATHEMATICAL PROBLEMS

2- PROBLEM SOLVING

3- DETERMINING THE ACTUAL SIZE OF AN OBJECT OBSERVED WITH A MICROSCOPE

4- DETERMINING THE MAGNIFICATION FACTOR FOR A DRAWING

5- DOING CONVERSIONS.

YOU ARE TO ANSWER THE FOLLOWING QUESTIONS AND SOLVE THE FOLLOWING PROBLEMS.

SHOW HOW YOU SOLVED THE PROBLEMS.

1. WHAT IS THE FORMULA USED TO CALCULATE THE VALUE IN UM FOR 1 OCULAR MICROMETER UNIT (SPACE)?

2. WHAT IS THE VALUE IN UM OF ONE UNIT (SPACE) OF AN OCULAR MICROMETER IF 15 STAGE MICROMETER UNITS EQUAL 45 OCULAR UNITS AND EACH STAGE MICROMETER UNIT IS EQUAL TO 2 MM?

3. IF AN OBJECT MEASURES 15 OCULAR MICROMETER UNITS (SPACES) AND EACH OCULAR MICROMETER UNIT HAS A VALUE OF 0.5 UM, HOW LARGE IS THE OBJECT IN UM?

4. IF AN OBJECT MEASURES 27 OCULAR MICROMETER UNITS AND EACH OCULAR MICROMETER UNIT HAS A VALUE OF 1 MM, HOW LARGE IS THE OBJECT IN UM?

5. WHAT IS THE FORMULA FOR CALCULATING THE MAGNIFICATION FACTOR OF A DRAWING?

6. WHAT IS THE MAGNIFICATION FACTOR FOR A DRAWING IF THE ACTUAL SIZE OF THE OBJECT IS 10 UM AND THE SIZE OF THE DRAWING IS 4 CM?

7. WHAT IS THE MAGNIFICATION FACTOR FOR A DRAWING IF THE ACTUAL SIZE OF THE OBJECT IS 1.5 UM AND THE SIZE OF THE DRAWING IS 6 CM?

8. WHAT IS THE FORMULA FOR DETERMINING THE ACTUAL SIZE OF AN OBJECT FROM A MICROGRAPH?

9. WHAT IS THE ACTUAL SIZE OF AN OBJECT IN UM IF THE TOTAL MAGNIFICATION IS 15,000x (15K) AND THE OBJECT MEASURES 2 CM IN THE PHOTOMICROGRAPH?

10. IF THE TOTAL MAGNIFICATION OF THE PHOTOMICROGRAPH IS 27,000X (27K) AND THE OBJECT IN THE PHOTOMICROGRAPH MEASURES 9 MM, WHAT IS THE ACTUAL SIZE OF THE OBJECT IN UM?

ASSESSMENT:

THE ASSIGNMENT IS DUE NEXT WEEK.
THE ASSIGNMENT IS WORTH TEN POINTS AND WILL BE ASSESSED ON CORRECTNESS OF METHODOLOGY AS WELL AS THE CORRECTNESS OF THE ANSWER.

THE IMPLEMENTATION:

THIS ASSIGNMENT IS HANDED OUT AT THE END OF THE LAB SESSION AND DISCUSSED WITH THE CLASS.

ASSIGNMENT # 8

GENERAL BIOLOGY I

LABORATORY EXERCISE #2   PHYSICAL PROCESSES: DIFFUSION

THE ASSIGNMENT:

THE PURPOSE OF THIS ASSIGNMENT IS TO GIVE YOU EXPERIENCE IN GRAPHICALLY PRESENTING DATA AND IN DRAWING INFERENCES FROM THE DATA.

USING THE DATA YOU COLLECTED FOR THE EXPERIMENT ON THE EFFECT OF MOLECULAR WEIGHT ON DIFFUSION RATES THROUGH A COLLOID, YOU ARE TO PREPARE TWO GRAPHS AND THEN MAKE SOME INFERENCES FROM THE DATA.

FOR THE GRAPHS BE SURE TO:

1- LABEL BOTH OF THE AXES OF THE GRAPH

2- GIVE AN INFORMATIVE TITLE OR CAPTION (REREAD PECHENIK CHAPTER 2 ON GRAPHING)

3- INCLUDE A KEY (IF NECESSARY)

4- PRESENT EACH GRAPH ON A SEPARATE PIECE OF GRAPH PAPER.

GRAPH I

MAKE A GRAPH FOR YOUR DATA FOR ALL KNOWN AND UNKNOWN COMPOUNDS USING DISTANCE (MM) RELATIVE TO TIME(MIN). CONNECT THE GRAPHED POINTS FOR EACH COMPOUND. THERE WILL BE EIGHT LINES ON THE GRAPH. MAKE A KEY TO IDENTIFY THE LINES.

GRAPH II

MAKE A GRAPH OF YOUR RESULTS FOR EACH KNOWN COMPOUND USING TOTAL DISTANCE DIFFUSED (MM) RELATIVE TO MOLECULAR WEIGHT (MOL.WT.). DRAW A SMOOTH CURVE TO CLARIFY THE TREND INDICATED BY YOUR GRAPHED POINTS. NEXT PLOT THE DATA FOR THE UNKNOWN COMPOUNDS ON THE SMOOTH CURVE AND DETERMINE THEIR MOLECULAR WEIGHTS. MAKE A KEY FOR THE UNKNOWNS AND THEIR DETERMINED MOLECULAR WEIGHTS.

YOUR LAB PARTNER HAS JUST PHONED AND WANTS TO KNOW WHAT YOUR GRAPH II LOOKS LIKE. DESCRIBE YOUR GRAPH FOR THEM AND THEN DISCUSS WHAT RELATIONSHIP IS INDICATED BETWEEN MOLECULAR WEIGHT AND THE RATE OF DIFFUSION. THIS PART OF THE ASSIGNMENT SHOULD BE ABOUT 150 WORDS OR 1/2 A PAGE. THIS MAY BE TYPED DOUBLE SPACED OR WRITTEN LEGIBLY.

ASSESSMENT:

THE TWO GRAPHS, YOUR DESCRIPTION AND DISCUSSION OF GRAPH II ARE DUE AT THE NEXT LAB SESSION.
THE ASSIGNMENT IS WORTH TEN POINTS. LATE ASSIGNMENTS RECEIVE LOWER GRADES.
THE ASSIGNMENT WILL BE ASSESSED ON THE QUALITY AND COMPLETENESS OF THE ASSIGNMENT.

THE IMPLEMENTATION:

I HAND OUT THE ASSIGNMENT AND DISCUSS IT AT THE BEGINNING OF THE LAB SESSION.

ASSIGNMENT # 9

GENERAL BIOLOGY I

LABORATORY EXERCISE #3   Saccaharides

THE ASSIGNMENT:

This assignments has two parts.

PART I

The purpose of this part of the assignment is to give you practice in presenting qualitative data in a visual and more understandable form for you and your lab partner.

You are to graphically present the results of the Benedict's test from page 17. Be sure to label both axes, give an informative caption and a key. Refer to Pechenik chapter 2.

PART II

Using the lab report format below (Assignment #10), write up the exercise on enzymatic hydrolysis: amylase action on starch.

The purpose of this assignment is to help you to better understand what you are doing in laboratory by making the numerous steps clearly separate. You will be practicing your observational skills, identifying processes, identifying assumptions, and making inferences or drawing conclusions.

Basic vocabulary:

1- observe: to see or sense through directed, careful, analytic attention

2- process: what you actually do or did

3- assumption: what you take for granted

4- infer: to arrive at a conclusion from facts and/or premise e.g. you see smoke and you infer the presence of fire

5- conclude: a reasoned judgment or inference

BE SPECIFIC IN YOU ANSWERS.

LABORATORY REPORT FORMAT

Answer all questions.
Questions 1-7 should be thought about BEFORE you do the

EXPERIMENT OR EXERCISE.

1- WHAT REACTION(S) OR PROCESS(ES) ARE YOU WORKING WITH OR INVESTIGATING? GIVE THE REACTION(S) OR PROCESS(ES) IN FOLLOWING FORM:

REACTANTS AND PRODUCT(S)   (A + B --> C + D)
EXAMPLE: SUCROSE(SUBSTRATE) + SUCRASE(ENZYME) --> GLUCOSE AND FRUCTOSE (PRODUCTS)

2- A. HOW ARE YOU TESTING FOR THE REACTION(S) OR WHAT TEST(S) ARE YOU CARRYING OUT TO SEE IF THE REACTION(S) OCCURRED?

B. WHAT REAGENT(S) ARE YOU USING?

C. HOW IS THE TEST PROCEDURE DONE?

D. WHAT DO YOU EXPECT TO OBSERVE FOR A POSITIVE TEST?

E. WHAT CAN YOU INFER FROM A POSITIVE TEST RESULT?

F. WHAT DO YOU EXPECT TO OBSERVE FOR A NEGATIVE TEST?

G. WHAT CAN YOU INFER FROM A NEGATIVE TEST RESULT?

3- WHAT ASSUMPTIONS ARE YOU MAKING -
A. ABOUT THE REACTANT(S)?

B. ABOUT THE ENZYME (IF PRESENT)?

C. ABOUT THE PRODUCT(S)?

D. ABOUT THE TEST REAGENT(S)?

4- WHAT ARE THE VARIABLES IN THE EXPERIMENT OR EXERCISE?

5- WHAT IS THE CONTROL (IF THERE IS ONE)?

6- WHICH SPECIFIC ASSUMPTION(S) FROM #3 ARE YOU ACTUALLY TESTING IN THIS EXPERIMENT OR EXERCISE?

7- WRITE AN EXPLANATORY HYPOTHESIS FOR THIS EXPERIMENT OR EXERCISE. (WHAT IS THE PURPOSE?) BE SPECIFIC.

8- EXACTLY WHAT DID YOU OBSERVE DURING THE EXPERIMENT OR EXERCISE?

OBSERVATIONS SHOULD BE NOTED AND RECORDED DURING THE ENTIRE PROCEDURE NOT JUST AT THE END. GIVE YOUR OBSERVATIONS FOR EACH SITUATION (TEST TUBE). THIS MAY BE A DATA SHEET WITH ADDITIONAL COMMENTS (ATTACH YOUR WORK).

9- GRAPH OR VISUALLY PRESENT YOUR OBSERVED/RECORDED RESULTS (ATTACH YOUR WORK).

10- WHAT INFERENCES CAN YOU MAKE FOR EACH TEST SITUATION OR TEST TUBE FROM YOUR OBSERVATIONS?

11- GRAPH OR VISUALLY PRESENT THE RESULTS YOU EXPECTED (ATTACH YOUR WORK).

12- WHY DID YOU EXPECT THE RESULTS IN #11? (WHAT WAS YOU REASONING?)

13- WHAT OVERALL OR SUMMARY CONCLUSION CAN YOU MAKE FOR THE WHOLE EXPERIMENT OR EXERCISE?

14- DID YOUR EXPERIMENTAL RESULTS AGREE WITH THE RESULTS YOU EXPECTED? IF THEY DID NOT AGREE, GIVE YOUR EXPLANATION FOR THE DIFFERENCES.

ASSESSMENT:

PART I

THIS IS DUE AT THE BEGINNING OF THE NEXT LABORATORY SESSION. THIS PART OF THE ASSIGNMENT IS WORTH 10 POINTS.
ASSESSMENT WILL BE BASED ON COMPLETENESS AND QUALITY OF THE ASSIGNMENT.

PART II

YOUR ROUGH DRAFT IS DUE AT THE BEGINNING OF THE NEXT LABORATORY SESSION.
AT THE BEGINNING OF THE NEXT LAB SESSION YOU AND YOUR LAB TEAM WILL DISCUSS YOUR ROUGH DRAFT AND RECORD ANY SUGGESTIONS ON THE BACK OF THE ROUGH DRAFT. TEAM MEMBERS SHOULD MAKE SPECIFIC SUGGESTIONS ON AREAS THAT NEED TO BE CLARIFIED. THIS IS AN OPPORTUNITY FOR YOU TO DISCOVER WHERE YOU HAVE QUESTIONS AND WHERE YOUR ROUGH DRAFT CAN BE IMPROVED.
YOUR FINAL LAB REPORT AND ROUGH DRAFT ARE DUE AT THE BEGINNING OF THE NEXT LABORATORY SESSION.
THIS PART OF THE ASSIGNMENT IS WORTH 20 POINTS AND WILL BE ASSESSED ON COMPLETENESS AND QUALITY.

THE IMPLEMENTATION:

WEEK 1- I HAND OUT THE ASSIGNMENT AND DISCUSS IT AT THE BEGINNING OF THE LAB SESSION.
WEEK 2- I COLLECT PART I, AND THE TEAMS GO OVER AND RECORD THEIR COMMENTS ON PART II.
WEEK 3- I COLLECT THE ROUGH DRAFTS AND FINAL LAB REPORTS FOR PART II.

ASSIGNMENT # 10

GENERAL BIOLOGY I

LABORATORY EXERCISE LABORATORY REPORT FORMAT

THE ASSIGNMENT:

ALTHOUGH YOU WILL BE HANDING IN ONLY A FEW LAB REPORTS FOR ASSESSMENT DURING THE SEMESTER, YOU SHOULD BE ABLE TO ANSWER THE FOLLOWING QUESTIONS FOR EVERY LABORATORY EXPERIMENT OR EXERCISE YOU DO. THE PURPOSE OF THIS ASSIGNMENT IS TO HELP YOU TO BETTER UNDERSTAND WHAT YOU ARE DOING IN LABORATORY BY MAKING THE NUMEROUS STEPS CLEARLY SEPARATE. YOU WILL BE PRACTICING YOUR OBSERVATIONAL SKILLS, IDENTIFYING PROCESSES, IDENTIFYING ASSUMPTIONS, AND MAKING INFERENCES OR DRAWING CONCLUSIONS.

BASIC VOCABULARY:

1- OBSERVE: TO SEE OR SENSE THROUGH DIRECTED, CAREFUL, ANALYTIC ATTENTION

2- PROCESS: WHAT YOU ACTUALLY DO OR DID

3- ASSUMPTION: WHAT YOU TAKE FOR GRANTED

4- INFER: TO ARRIVE AT A CONCLUSION FROM FACTS AND/OR PREMISE E.G. YOU SEE SMOKE AND YOU INFER THE PRESENCE OF FIRE

5- CONCLUDE: A REASONED JUDGMENT OR INFERENCE

BE SPECIFIC IN YOU ANSWERS.

LABORATORY REPORT FORMAT

ANSWER ALL QUESTIONS.
QUESTIONS 1-7 SHOULD BE THOUGHT ABOUT BEFORE YOU DO THE EXPERIMENT OR EXERCISE.

1- WHAT REACTION(S) OR PROCESS(ES) ARE YOU WORKING WITH OR INVESTIGATING? GIVE THE REACTION(S) OR PROCESS(ES) IN FOLLOWING FORM:

REACTANTS AND PRODUCT(S)   (A + B --> C + D)
EXAMPLE: SUCROSE(SUBSTRATE) + SUCRASE(ENZYME) --> GLUCOSE AND FRUCTOSE (PRODUCTS)

2- A. HOW ARE YOU TESTING FOR THE REACTION(S) OR WHAT TEST(S) ARE YOU CARRYING OUT TO SEE IF THE REACTION(S) OCCURRED?

B. WHAT REAGENT(S) ARE YOU USING?

C. HOW IS THE TEST PROCEDURE DONE?

D. WHAT DO YOU EXPECT TO OBSERVE FOR A POSITIVE TEST?

E. WHAT CAN YOU INFER FROM A POSITIVE TEST RESULT?

F. WHAT DO YOU EXPECT TO OBSERVE FOR A NEGATIVE TEST?

G. WHAT CAN YOU INFER FROM A NEGATIVE TEST RESULT?

3- WHAT ASSUMPTIONS ARE YOU MAKING -
A. ABOUT THE REACTANT(S)?

B. ABOUT THE ENZYME (IF PRESENT)?

C. ABOUT THE PRODUCT(S)?

D. ABOUT THE TEST REAGENT(S)?

4- WHAT ARE THE VARIABLES IN THE EXPERIMENT OR EXERCISE?

5- WHAT IS THE CONTROL (IF THERE IS ONE)?

6- WHICH SPECIFIC ASSUMPTION(S) FROM #3 ARE YOU ACTUALLY TESTING IN THIS EXPERIMENT OR EXERCISE?

7- WRITE AN EXPLANATORY HYPOTHESIS FOR THIS EXPERIMENT OR EXERCISE. (WHAT IS THE PURPOSE?) BE SPECIFIC.

8- EXACTLY WHAT DID YOU OBSERVE DURING THE EXPERIMENT OR EXERCISE?

OBSERVATIONS SHOULD BE NOTED AND RECORDED DURING THE ENTIRE PROCEDURE NOT JUST AT THE END. GIVE YOUR OBSERVATIONS FOR EACH SITUATION (TEST TUBE). THIS MAY BE A DATA SHEET WITH ADDITIONAL COMMENTS (ATTACH YOUR WORK).

9- GRAPH OR VISUALLY PRESENT YOUR OBSERVED/RECORDED RESULTS (ATTACH YOUR WORK).

10- WHAT INFERENCES CAN YOU MAKE FOR EACH TEST SITUATION OR TEST TUBE FROM YOUR OBSERVATIONS?

11- GRAPH OR VISUALLY PRESENT THE RESULTS YOU EXPECTED (ATTACH YOUR WORK).

12- WHY DID YOU EXPECT THE RESULTS IN #11? (WHAT WAS YOU REASONING?)

13- WHAT OVERALL OR SUMMARY CONCLUSION CAN YOU MAKE FOR THE WHOLE EXPERIMENT OR EXERCISE?

14- Did your experimental results agree with the results you expected? If they did not agree, give your explanation for the differences.

THE IMPLEMENTATION:

I hand out this assignment at the beginning of the semester and discuss it with the class. I incorporate the lab report format in those assignments that involve lab reports.

ASSIGNMENT # 11

GENERAL BIOLOGY I

LABORATORY EXERCISE #5 ENZYMES

THE ASSIGNMENT:

USING THE LAB REPORT FORMAT BELOW (ASSIGNMENT #10), WRITE UP THE EXERCISE: EXPERIMENTAL PROCEDURE - TIME COURSE OF ENZYMATIC REACTIONS. USE THE RESULTS FROM BOTH THE 3% AND 10% SUCROSE EXPERIMENTS.

THE PURPOSE OF THIS ASSIGNMENT IS TO HELP YOU TO BETTER UNDERSTAND WHAT YOU ARE DOING IN LABORATORY BY MAKING THE NUMEROUS STEPS CLEARLY SEPARATE. YOU WILL BE PRACTICING YOUR OBSERVATIONAL SKILLS, IDENTIFYING PROCESSES, IDENTIFYING ASSUMPTIONS, AND MAKING INFERENCES OR DRAWING CONCLUSIONS.

BASIC VOCABULARY:

1- OBSERVE: TO SEE OR SENSE THROUGH DIRECTED, CAREFUL, ANALYTIC ATTENTION

2- PROCESS: WHAT YOU ACTUALLY DO OR DID

3- ASSUMPTION: WHAT YOU TAKE FOR GRANTED

4- INFER: TO ARRIVE AT A CONCLUSION FROM FACTS AND/OR PREMISE E.G. YOU SEE SMOKE AND YOU INFER THE PRESENCE OF FIRE

5- CONCLUDE: A REASONED JUDGMENT OR INFERENCE

BE SPECIFIC IN YOU ANSWERS.

LABORATORY REPORT FORMAT

ANSWER ALL QUESTIONS.
QUESTIONS 1-7 SHOULD BE THOUGHT ABOUT BEFORE YOU DO THE EXPERIMENT OR EXERCISE.

1- WHAT REACTION(S) OR PROCESS(ES) ARE YOU WORKING WITH OR INVESTIGATING? GIVE THE REACTION(S) OR PROCESS(ES) IN FOLLOWING FORM:

REACTANTS AND PRODUCT(S)   (A + B --> C + D)
EXAMPLE: SUCROSE(SUBSTRATE) + SUCRASE(ENZYME) --> GLUCOSE AND FRUCTOSE (PRODUCTS)

2- A. HOW ARE YOU TESTING FOR THE REACTION(S) OR WHAT TEST(S) ARE YOU CARRYING OUT TO SEE IF THE REACTION(S) OCCURRED?

B. WHAT REAGENT(S) ARE YOU USING?

C. HOW IS THE TEST PROCEDURE DONE?

D. WHAT DO YOU EXPECT TO OBSERVE FOR A POSITIVE TEST?

E. WHAT CAN YOU INFER FROM A POSITIVE TEST RESULT?

F. WHAT DO YOU EXPECT TO OBSERVE FOR A NEGATIVE TEST?

G. WHAT CAN YOU INFER FROM A NEGATIVE TEST RESULT?

3- WHAT ASSUMPTIONS ARE YOU MAKING -
A. ABOUT THE REACTANT(S)?

B. ABOUT THE ENZYME (IF PRESENT)?

C. ABOUT THE PRODUCT(S)?

D. ABOUT THE TEST REAGENT(S)?

4- WHAT ARE THE VARIABLES IN THE EXPERIMENT OR EXERCISE?

5- WHAT IS THE CONTROL (IF THERE IS ONE)?

6- WHICH SPECIFIC ASSUMPTION(S) FROM #3 ARE YOU ACTUALLY TESTING IN THIS EXPERIMENT OR EXERCISE?

7- WRITE AN EXPLANATORY HYPOTHESIS FOR THIS EXPERIMENT OR EXERCISE. (WHAT IS THE PURPOSE?) BE SPECIFIC.

8- EXACTLY WHAT DID YOU OBSERVE DURING THE EXPERIMENT OR EXERCISE?

OBSERVATIONS SHOULD BE NOTED AND RECORDED DURING THE ENTIRE PROCEDURE NOT JUST AT THE END. GIVE YOUR OBSERVATIONS FOR EACH SITUATION (TEST TUBE). THIS MAY BE A DATA SHEET WITH ADDITIONAL COMMENTS (ATTACH YOUR WORK).

9- GRAPH OR VISUALLY PRESENT YOUR OBSERVED/RECORDED RESULTS (ATTACH YOUR WORK).

10- WHAT INFERENCES CAN YOU MAKE FOR EACH TEST SITUATION OR TEST TUBE FROM YOUR OBSERVATIONS?

11- GRAPH OR VISUALLY PRESENT THE RESULTS YOU EXPECTED (ATTACH YOUR WORK).

12- WHY DID YOU EXPECT THE RESULTS IN #11? (WHAT WAS YOU REASONING?)

13- WHAT OVERALL OR SUMMARY CONCLUSION CAN YOU MAKE FOR THE WHOLE

EXPERIMENT OR EXERCISE?

14- DID YOUR EXPERIMENTAL RESULTS AGREE WITH THE RESULTS YOU EXPECTED? IF THEY DID NOT AGREE, GIVE YOUR EXPLANATION FOR THE DIFFERENCES.

ASSESSMENT:

YOUR ROUGH DRAFT IS DUE AT THE BEGINNING OF THE NEXT LABORATORY SESSION.

AT THE BEGINNING OF THE NEXT LAB SESSION YOU AND YOUR LAB TEAM WILL DISCUSS YOUR ROUGH DRAFT AND RECORD ANY SUGGESTIONS ON THE BACK OF THE ROUGH DRAFT. TEAM MEMBERS SHOULD MAKE SPECIFIC SUGGESTIONS ON AREAS THAT NEED TO BE CLARIFIED. THIS IS AN OPPORTUNITY FOR YOU TO DISCOVER WHERE YOU HAVE QUESTIONS AND WHERE YOUR ROUGH DRAFT CAN BE IMPROVED.

YOUR FINAL LAB REPORT ALONG WITH YOUR ROUGH DRAFT ARE DUE AT THE BEGINNING OF THE NEXT LAB SESSION.

THIS ASSIGNMENT IS WORTH 30 POINTS AND WILL BE ASSESSED ON COMPLETENESS AND QUALITY.

THE IMPLEMENTATION:

WEEK 1- I HAND OUT THE ASSIGNMENT AND DISCUSS IT AT THE BEGINNING OF THE LAB SESSION.

WEEK 2- THE TEAMS GO OVER THE ROUGH DRAFTS AND RECORD THEIR COMMENTS.

WEEK 3- I COLLECT THE ROUGH DRAFTS AND FINAL LAB REPORTS.

ASSIGNMENT # 12

GENERAL BIOLOGY I

LABORATORY EXERCISE #6   PHOTOSYNTHESIS

THE ASSIGNMENT:

THE PURPOSE OF THIS ASSIGNMENT IS TO GIVE YOU PRACTICE IN ORGANIZING THE COLLECTION OF, THE GRAPHIC REPRESENTATION OF, AND THE ANALYSIS OF DATA. YOU WILL ALSO BE PRACTICING YOUR ABILITY TO ANALYZE INFORMATION AND DRAW CONCLUSIONS.

THIS ASSIGNMENT HAS FOUR PARTS.

PART I

TO SHOW ME HOW YOU ORGANIZED THE GATHERING OF YOUR DATA FOR THE PHOTOSYNTHETIC PIGMENT EXPERIMENT, HAND IN YOUR DATA SHEET.

PART II

PRESENT GRAPHICALLY IN A MEANINGFUL WAY YOUR QUANTITATIVE DATA FROM THE PHOTOSYNTHETIC PIGMENT EXPERIMENT. PLACE THE DATA FOR ALL SOLUTIONS ON ONE GRAPH. BE SURE TO LABEL BOTH AXES, DESIGN A KEY, AND GIVE AN INFORMATIVE CAPTION (REFER TO PECHENIK CHAPTER 2).

PART III

FROM YOUR GRAPH DETERMINE AT WHICH WAVELENGTH(S) EACH PIGMENT ABSORBED THE MOST. PRESENT THIS INFORMATION AS A TABLE. BASED ON THE INFORMATION IN YOUR TABLE, WHICH PIGMENT(S) DO YOU CONCLUDE WOULD CONTRIBUTE THE MOST TO THE PROCESS OF PHOTOSYNTHESIS?

PART IV

YOUR PARTNERS WERE OUT SICK THE DAY THIS EXPERIMENT WAS DONE. TO HELP THEM UNDERSTAND YOUR RESULTS MORE FULLY, COMPARE (SIMILARITIES AND DIFFERENCES) YOUR GRAPH (ABSORPTION SPECTRA) WITH THE ONE ON PAGE ___ IN YOUR TEXT BOOK. BE SURE TO EXPLAIN HOW ANY DISCREPANCIES BETWEEN YOUR SPECTRUM AND THE PUBLISHED SPECTRUM MAY HAVE OCCURRED. THIS PART OF THE ASSIGNMENT SHOULD BE ABOUT ONE PAGE (APPROXIMATELY 250 WORDS) IN LENGTH.

ASSESSMENT:

ALL FOUR PARTS ARE DUE AT THE BEGINNING OF THE NEXT LABORATORY SESSION.
THIS ASSIGNMENTS IS WORTH 20 POINTS.
THE ASSIGNMENT WILL BE ASSESSED ON THE COMPLETENESS AND

QUALITY OF THE WORK.

THE IMPLEMENTATION:

THE ASSIGNMENT IS HANDED OUT AND DISCUSSED AT THE BEGINNING OF THE LAB SESSION.

ASSIGNMENT # 13

GENERAL BIOLOGY I

LABORATORY EXERCISE #6 RESPIRATION

THE ASSIGNMENT:

USING THE LAB REPORT FORMAT BELOW (ASSIGNMENT #10), WRITE UP THE EXERCISE: EXPERIMENTAL PROCEDURE - THE EFFECT OF MAGNESIUM AND FLUORIDE ON ALCOHOLIC FERMENTATION.

THE PURPOSE OF THIS ASSIGNMENT IS TO HELP YOU TO BETTER UNDERSTAND WHAT YOU ARE DOING IN LABORATORY BY MAKING THE NUMEROUS STEPS CLEARLY SEPARATE. YOU WILL BE PRACTICING YOUR OBSERVATIONAL SKILLS, IDENTIFYING PROCESSES, IDENTIFYING ASSUMPTIONS, AND MAKING INFERENCES OR DRAWING CONCLUSIONS.

BASIC VOCABULARY:

1- OBSERVE: TO SEE OR SENSE THROUGH DIRECTED, CAREFUL, ANALYTIC ATTENTION

2- PROCESS: WHAT YOU ACTUALLY DO OR DID

3- ASSUMPTION: WHAT YOU TAKE FOR GRANTED

4- INFER: TO ARRIVE AT A CONCLUSION FROM FACTS AND/OR PREMISE E.G. YOU SEE SMOKE AND YOU INFER THE PRESENCE OF FIRE

5- CONCLUDE: A REASONED JUDGMENT OR INFERENCE

BE SPECIFIC IN YOU ANSWERS.

LABORATORY REPORT FORMAT

ANSWER ALL QUESTIONS.
QUESTIONS 1-7 SHOULD BE THOUGHT ABOUT BEFORE YOU DO THE EXPERIMENT OR EXERCISE.

1- WHAT REACTION(S) OR PROCESS(ES) ARE YOU WORKING WITH OR INVESTIGATING? GIVE THE REACTION(S) OR PROCESS(ES) IN FOLLOWING FORM:

REACTANTS AND PRODUCT(S)   (A + B --> C + D)
EXAMPLE: SUCROSE(SUBSTRATE) + SUCRASE(ENZYME) --> GLUCOSE AND FRUCTOSE (PRODUCTS)

2- A. HOW ARE YOU TESTING FOR THE REACTION(S) OR WHAT TEST(S) ARE YOU CARRYING OUT TO SEE IF THE REACTION(S) OCCURRED?

B. WHAT REAGENT(S) ARE YOU USING?

C. HOW IS THE TEST PROCEDURE DONE?

D. WHAT DO YOU EXPECT TO OBSERVE FOR A POSITIVE TEST?

E. WHAT CAN YOU INFER FROM A POSITIVE TEST RESULT?

F. WHAT DO YOU EXPECT TO OBSERVE FOR A NEGATIVE TEST?

G. WHAT CAN YOU INFER FROM A NEGATIVE TEST RESULT?

3- WHAT ASSUMPTIONS ARE YOU MAKING -
A. ABOUT THE REACTANT(S)?

B. ABOUT THE ENZYME (IF PRESENT)?

C. ABOUT THE PRODUCT(S)?

D. ABOUT THE TEST REAGENT(S)?

4- WHAT ARE THE VARIABLES IN THE EXPERIMENT OR EXERCISE?

5- WHAT IS THE CONTROL (IF THERE IS ONE)?

6- WHICH SPECIFIC ASSUMPTION(S) FROM #3 ARE YOU ACTUALLY TESTING
IN THIS EXPERIMENT OR EXERCISE?

7- WRITE AN EXPLANATORY HYPOTHESIS FOR THIS EXPERIMENT OR
EXERCISE. (WHAT IS THE PURPOSE?) BE SPECIFIC.

8- EXACTLY WHAT DID YOU OBSERVE DURING THE EXPERIMENT OR
EXERCISE?

OBSERVATIONS SHOULD BE NOTED AND RECORDED DURING THE ENTIRE
PROCEDURE NOT JUST AT THE END. GIVE YOUR OBSERVATIONS FOR
EACH SITUATION (TEST TUBE). THIS MAY BE A DATA SHEET WITH
ADDITIONAL COMMENTS (ATTACH YOUR WORK).

9- GRAPH OR VISUALLY PRESENT YOUR OBSERVED/RECORDED RESULTS
(ATTACH YOUR WORK).

10- WHAT INFERENCES CAN YOU MAKE FOR EACH TEST SITUATION OR TEST
TUBE FROM YOUR OBSERVATIONS?

11- GRAPH OR VISUALLY PRESENT THE RESULTS YOU EXPECTED (ATTACH
YOUR WORK).

12- WHY DID YOU EXPECT THE RESULTS IN #11? (WHAT WAS YOU
REASONING?)

13- WHAT OVERALL OR SUMMARY CONCLUSION CAN YOU MAKE FOR THE WHOLE

EXPERIMENT OR EXERCISE?

14- DID YOUR EXPERIMENTAL RESULTS AGREE WITH THE RESULTS YOU EXPECTED? IF THEY DID NOT AGREE, GIVE YOUR EXPLANATION FOR THE DIFFERENCES.

ASSESSMENT:

YOUR ROUGH DRAFT IS DUE AT THE NEXT LABORATORY ASESSION.
AT THE BEGINNING OF THE NEXT LAB SESSION YOU AND YOUR LAB TEAM WILL DISCUSS YOUR ROUGH DRAFT AND RECORD ANY SUGGESTIONS ON THE BACK OF THE ROUGH DRAFT. TEAM MEMBERS SHOULD MAKE SPECIFIC SUGGESTIONS ON AREAS THAT NEED TO BE CLARIFIED. THIS AN OPPORTUNITY FOR YOU TO DISCOVER WHERE YOU HAVE QUESTIONS AND WHERE YOUR ROUGH DRAFT CAN BE IMPROVED.
YOUR FINAL LAB REPORT ALONG WITH YOUR ROUGH DRAFT ARE DUE AT THE BEGINNING OF THE NEXT LAB SESSION.
THIS ASSIGNMENT IS WORTH 30 POINTS AND WILL BE ASSESSED ON COMPLETENESS AND QUALITY.

THE IMPLEMENTATION:

WEEK 1- I HAND OUT THE ASSIGNMENT AND DISCUSS IT AT THE BEGINNING OF THE LAB SESSION.
WEEK 2- THE TEAMS GO OVER THE ROUGH DRAFTS AND RECORD THEIR COMMENTS.
WEEK 3- I COLLECT THE ROUGH DRAFTS AND FINAL LAB REPORTS.

ASSIGNMENT # 14

GENERAL BIOLOGY I

LABORATORY EXERCISE #9   MITOSIS AND MEIOSIS

THE ASSIGNMENT:

THIS ASSIGNMENT IS DESIGNED TO HELP YOU SHARPEN YOUR OBSERVATIONAL AND RECORDING SKILLS. YOU WILL ALSO GET MORE PRACTICE USING THE OCULAR MICROMETER.

BEFORE YOU START YOUR MICROSCOPE WORK LOOK AT ALL CLASSROOM MODELS AND CHARTS SO YOU CAN IDENTIFY THE VARIOUS PHASES OF THE CONTINUOUS PROCESSES OF MITOSIS AND MEIOSIS.

MAKE SKETCHES OF ALL THE MITOTIC PHASES YOU OBSERVE ON THE ROOT TIP AND WHITEFISH BLASTULA SLIDES.

YOU ARE TO MAKE TWO DRAWINGS TO HAND IN AT THE NEXT LAB SESSION. BOTH DRAWINGS MUST BE OF THE SAME PHASE OF MITOSIS, EITHER CELLS IN METAPHASE OR IN ANAPHASE. DRAW A CELL FROM THE ROOT TIP SLIDE AND ONE FROM THE WHITEFISH BLASTULA SLIDE FOLLOWING THE DRAWING FORMAT HANDED OUT AT THE FIRST LABORATORY SESSION. THE COMPARISON FOR THESE TWO DRAWINGS WILL BE BETWEEN THE PLANT CELL AND THE ANIMAL CELL.

ASSESSMENT:

YOUR TWO DRAWINGS ARE DUE AT THE BEGINNING OF THE NEXT LAB SESSION.
THE DRAWINGS ARE WORTH TEN POINTS.
THE DRAWINGS WILL BE ASSESSED ON THE COMPLETENESS OF THE INFORMATION PRESENTED AND ON THE CLARITY AND ACCURACY OF THE DRAWINGS (THIS DOES NOT MEAN ARTISTIC ABILITY).

THE IMPLEMENTATION:

THE ASSIGNMENT IS HANDED OUT AT THE BEGINNING OF THE LAB SESSION AND DISCUSSED.

ASSIGNMENT # 15

GENERAL BIOLOGY I

LABORATORY EXERCISE #10 PATTERNS OF INHERITANCE: MENDELIAN
GENETICS

THE ASSIGNMENT:

THE PURPOSE OF THIS ASSIGNMENT IS TO GIVE YOU PRACTICE IN:
1- TRANSLATING WORD PROBLEMS INTO SYMBOL PROBLEMS AND SOLVING
THEM
2- USING THE PUNNETT SQUARE
3- COLLECTING, ORGANIZING AND ANALYZING DATA
4- DRAWING INFERENCES FROM YOUR DATA.

THIS ASSIGNMENT HAS TWO PARTS.

PART I

YOU AND YOUR LAB PARTNER WILL DO THE EXERCISE AS DESCRIBED IN
THE CORN DIHYBRID GENETICS INFORMATION SHEET HANDED OUT IN CLASS.
FIRST COMPLETE THE PUNNETT SQUARE SIDE OF THE HANDOUT. SECOND
COLLECT AND RECORD DATA FOR TWO EARS OF CORN. THIRD CALCULATE
THE EXPECTED NUMBERS FOR YOUR SAMPLE DATA BASED ON MENDEL'S
EXPECTED RATIOS FOR A DIHYBRID CROSS.

CLASS DATA FOR THIS EXERCISE WILL BE PUT ON THE BLACK BOARD.
AFTER YOU HAVE RECORDED THE CLASS DATA, CALCULATE THE EXPECTED
RESULTS FOR THE CLASS DATA AS YOU DID FOR YOUR OWN DATA.

PART II

A FRIEND IN ANOTHER LAB WAS ILL LAST WEEK AND YOU HAVE AGREED
TO HELP THEM UNDERSTAND THE RESULTS OF THIS EXERCISE, WRITE ONE
OR TWO PARAGRAPHS (@ 150-250 WORDS) INTERPRETING BOTH YOUR DATA
AND THE CLASS DATA. WHICH SET OF DATA WAS CLOSER TO MENDEL'S
EXPECTED RATIOS FOR A DIHYBRID CROSS? EXPLAIN WHICH SET OF DATA
YOU EXPECTED TO BE CLOSEST TO MENDEL'S RATIOS AND WHY.

ASSESSMENT:

BOTH PART I AND PART II ARE DUE AT THE BEGINNING OF THE NEXT
LAB SESSION.
THE ASSIGNMENT IS WORTH TEN POINTS.
PART I WILL BE ASSESSED ON THE COMPLETENESS AND CORRECTNESS
OF THE WORK. PART II WILL BE ASSESSED ON CLARITY AND
COMPLETENESS OF YOUR EXPLANATION.

THE IMPLEMENTATION:

THE ASSIGNMENT IS HANDED OUT AT THE BEGINNING OF THE LAB

SESSION AND DISCUSSED.

ASSIGNMENT # 16

GENERAL BIOLOGY I

LABORATORY EXERCISE #12 POPULATION GENETICS

THE ASSIGNMENT:

THE PURPOSE OF THIS ASSIGNMENT IS TO HELP YOU UNDERSTAND AND TO GIVE YOU PRACTICE IN USING THE HARDY-WEINBERG EQUATION. YOU WILL ALSO BE PRACTICING YOUR ABILITIES IN TRANSLATING WORD PROBLEMS INTO SYMBOL PROBLEMS, MAKING INFERENCES AND SUMMARIZING INFORMATION.

SHOW ALL OF YOUR WORK.
USE THE FOLLOWING EXAMPLE TO ANSWER QUESTIONS 1-11.

EXAMPLE: WITHIN A POPULATION OF RABBITS SOME HAVE LONG EARS (L) AND SOME HAVE SHORT EARS (L). OUT OF A POPULATION OF 10,000 RABBITS, 400 HAVE SHORT EARS.

QUESTIONS:

1. WHAT GENE IS BEING STUDIED?

2. WHAT ARE THE ALLELES OF THE GENE AND THEIR SYMBOLS?

3. WHAT IS THE HARDY-WEINBERG EQUATION?

4. WHAT IS THE FREQUENCY FOR EACH ALLELE?

5. HOW MANY RABBITS WILL BE HOMOZYGOUS DOMINANT FOR THE TRAIT?

6. HOW MANY RABBITS WILL BE HOMOZYGOUS RECESSIVE FOR THE TRAIT?

7. HOW MANY RABBITS WILL BE HETEROZYGOUS FOR THE TRAIT?

8. HOW MANY RABBITS WILL HAVE THE LONG EAR PHENOTYPE?

9. HOW MANY RABBITS WILL HAVE THE SHORT EAR PHENOTYPE?

10. IF YOU CAME BACK AND STUDIED THE SAME RABBIT POPULATION IN TEN YEARS AND FOUND THAT THE FREQUENCY OF THE LONG EAR ALLELE HAD INCREASED TWO TIMES WHAT IT IS NOW, WHAT INFERENCE COULD YOU MAKE ABOUT THE POPULATION? (ANSWER IN COMPLETE SENTENCES.)

11. BASED ON THE ABOVE EXAMPLE, EXPLAIN TO YOUR FAVORITE 15 YEAR OLD COUSIN HOW THE HARDY-WEINBERG EQUATION IS USED BY SCIENTISTS TO UNDERSTAND THE PROCESS OF EVOLUTION. YOUR

EXPLANATION SHOULD BE ABOUT 1/2 A PAGE OR 150 WORDS IN LENGTH EITHER TYPED OR WRITTEN.

ASSESSMENT:

THIS ASSIGNMENT IS DUE AT THE NEXT LAB SESSION AND IS WORTH 10 POINTS.

THIS ASSIGNMENT WILL BE ASSESSED ON COMPLETENESS AND DEMONSTRATED UNDERSTANDING OF THE HARDY-WEINBERG EQUATION AND ITS APPLICATION.

THE IMPLEMENTATION:

THIS ASSIGNMENT IS HANDED OUT AND DISCUSSED IN THE LAB SESSION.

ASSIGNMENT # 17

GENERAL BIOLOGY   II

LABORATORY EXERCISE   LABORATORY INTRODUCTION

THE ASSIGNMENT:

THIS IS THE FIRST LABORATORY EXERCISE OF THE SEMESTER.

1- EXPLAIN CLUSTERING USING A SPECIFIC EXAMPLE.

2- DISTRIBUTE PAPER AND HAVE THE STUDENTS INDIVIDUALLY CLUSTER ON THE TERM BIOLOGY LAB. THEN HAVE LAB PARTNERS EXCHANGE CLUSTERS SO THEY CAN SEE HOW ANOTHER PERSON VIEWS THE LABORATORY.

3- EXPLAIN FREE WRITING.

4- DISTRIBUTE PAPER AND HAVE THE STUDENTS FREE WRITE ON BIOLOGY LAB FOR 5 MIN.

5- DIVIDED THE CLASS INTO LAB TEAMS OF FOUR AND HAVE EACH MEMBER OF THE TEAM READ THEIR FREE WRITING ALOUD TO THE TEAM. AFTER GROUP DISCUSSION EACH TEAM CHOOSES ONE WRITING TO PRESENT TO THE CLASS.

6- LAB TEAM MEMBERS EXCHANGE NAMES AND PHONE NUMBERS WITH EACH OTHER AND ONE STUDENT, ACTING AS SECRETARY, MAKES A NAME/PHONE LIST FOR ME.

7- EACH TEAM READS THEIR CHOOSEN FREE WRITING ALOUD TO THE CLASS AND THE CLASS DISCUSSES THE FREE WRITING SAMPLES. THIS DISCUSSION LEADS INTO WHAT LAB IS GOING TO BE ABOUT THIS SEMESTER AND THE RULES OF THE ROAD. I HAND OUT INFORMATION ON THE LAB FORMAT, EVALUATIONS, JOURNALS, AND DRAWINGS.

THE IMPLEMENTATION:

ALL OF THE ASSIGNMENT IS DONE IN CLASS. AT THE END OF THE CLASS I COLLECT THE CLUSTERING AND FREE WRITING EXERCISES. I READ AND MAKE SUPPORTIVE COMMENTS ON THESE PAPERS AND RETURN THEM AT THE NEXT LAB SESSION.

ASSIGNMENT # 18

GENERAL BIOLOGY II

LABORATORY EXERCISE   CUBING EXERCISE

THE ASSIGNMENT:

THIS EXERCISE IS TO INTRODUCE YOU TO A TECHNIQUE WHICH WILL HELP YOU TO IMPROVE YOUR POWERS OF OBSERVATION. YOU WILL ALSO PRACTICE RECORDING YOU OBSERVATIONS BOTH VISUALLY AND IN WRITING. DURING THE EXERCISE YOU WILL HAVE THE CHANCE TO USE YOUR IMAGINATION.

READ THE ENTIRE EXERCISE BEFORE YOU START. YOU WILL NEED PENCIL, ERASER, DRAWING PAPER, WRITING PAPER, A LIGHT MICROSCOPE, AND A DISSECTING MICROSCOPE.

CUBING IS A TECHNIQUE FOR SWIFTLY CONSIDERING AN OBJECT FROM SEVERAL DIFFERENT POINTS OF VIEW.

IT IS IMPORTANT THAT YOU DO NOT DISCUSS YOUR WORK IN ANY WAY WITH YOUR PARTNER UNTIL BOTH OF YOUR HAVE FINISHED PART A.

THIS EXERCISE IS DONE WITH YOUR LAB PARTNER.

PART A:

PERSON A WILL GET A NUMBERED SLIDE OR OBJECT FROM THE FRONT OF THE ROOM AND WILL DO THE EXERCISE WHILE PERSON B READS THE QUESTIONS ALOUD TO AND KEEPS TIME FOR PERSON A.

AFTER PERSON A HAS FINISHED THE EXERCISE, A AND B WILL CHANGE JOBS AND REPEAT THE EXERCISE.

CUBING

DO EACH OF THE FOLLOWING STEPS IN ORDER, SPENDING NO MORE THAN 2 MINUTES ON EACH STEP. YOUR PARTNER'S JOBS ARE TO READ AND KEEP TIME. YOUR PARTNER WILL READ THE STEPS ALOUD TO YOU ONE AT A TIME AND TIME YOU FOR 2 MINUTES. YOU ARE TO LOOK AT THE SLIDE OR OBJECT AND WRITE DOWN YOUR RESPONSE TO THE QUESTION READ BY YOUR PARTNER. YOU ARE NOT TO READ THE QUESTION, YOUR JOB IS TO OBSERVE AND WRITE. YOUR RESPONSES ARE TO BE IN COMPLETE SENTENCES.

STEPS:

1- DESCRIBE IT. LOOK AT THE OBJECT CLOSELY AND DESCRIBE WHAT YOU SEE IN DETAIL (COLORS, SHAPES, SIZES, ETC).

2- COMPARE IT. WHAT IS IT SIMILAR TO? DIFFERENT FROM?

3- ASSOCIATE IT. WHAT DOES IT MAKE YOU THINK OF (TIMES, PLACES, PEOPLE, THINGS).

4- ANALYZE IT. TELL HOW YOU THINK IT WAS MADE. IF YOU DON'T KNOW MAKE UP YOUR ANSWER.

5- APPLY IT. DESCRIBE AND TELL WHAT YOU CAN DO WITH IT, HOW IT CAN BE USED.

6- ARGUE FOR OR AGAINST IT. TAKE A STAND - DO YOU LIKE IT OR NOT. GIVE A RATIONALE FOR YOUR STAND.

PART B:

AFTER YOU AND YOUR LAB PARTNER HAVE FINISHED PART A OF THE EXERCISE, YOU WILL COMPARE AND DISCUSS WHAT YOU HAVE DONE.

PART C:

AFTER YOU HAVE COMPLETED PARTS A AND B YOU ARE TO MAKE A DRAWING OF THE SLIDE OR OBJECT FOLLOWING THE DRAWING FORMAT HANDED OUT IN CLASS. REMEMBER TO DRAW WHAT YOU SEE NOT WHAT YOU THINK THE OBJECT IS SUPPOSED TO LOOK LIKE.

YOU ARE TO DO THE CUBING EXERCISE FOR TWO DIFFERENT SLIDES OR OBJECTS.

ASSESSMENT:

YOUR TWO CUBING EXERCISES AND DRAWINGS ARE DUE AT THE BEGINNING OF THE NEXT LAB SESSION.
CUBINGS AND DRAWINGS ARE WORTH A TOTAL OF TEN POINTS.
THIS ASSIGNMENT WILL BE ASSESSED ON THE COMPLETENESS OF YOUR WORK AND ON THE CLARITY OF YOUR DRAWINGS (CLARITY IS NOT ARTISTIC ABILITY).

THE IMPLEMENTATION:

THIS ASSIGNMENTS IS GIVEN OUT AND DISCUSSED AFTER THE TIPS AND HANDOUTS ON DRAWING HAVE BEEN GIVEN TO THE CLASS. I HAVE PREVIOUSLY PREPARED TWO TRAYS ONE WITH NUMBERED SLIDES AND ONE WITH LARGER SPECIMENS. THE SLIDES INCLUDE POLLEN, DIATOMS, FORAMINIFERA, AND PROTOZOAN. THE LARGER SPECIMENS ARE LICHENS, SEEDS, FRUITS, CORALS, SPONGES AND BONES. I TRY TO USE MATERIALS THAT WILL BE UNFAMILIAR TO THE STUDENTS.

ASSIGNMENT # 19

GENERAL BIOLOGY II

LABORATORY EXERCISE   MUSEUM TRIP AND REPORT

THE ASSIGNMENT:

YOU ARE TO VISIT THE AMERICAN MUSEUM OF NATURAL HISTORY ON CENTRAL PARK WEST AT 79TH STREET IN NEW YORK CITY. THE PURPOSE OF THIS ASSIGNMENT IS TO FAMILARIZE YOU WITH THE DIVERSE EXHIBITS OF THE MUSEUM, ESPECIALLY THOSE RELATED TO GENERAL BIOLOGY. AS YOU GO THROUGH THE VARIOUS EXHIBITS IN THE ROOMS LISTED BELOW YOU SHOULD BE AWARE OF THE EVOLUTIONARY RELATIONSHIPS EVIDENT IN MANY OF THE EXHIBITS.

YOU ARE BEING ASKED TO GO TO THE MUSEUM SO THAT YOU CAN:

1- EXPERIENCE BIOLOGY IN A DIFFERENT FRAMEWORK
2- ENJOY A NEW EXPERIENCE
3- REFLECT ON WHAT YOU HAVE SEEN, DONE, THOUGHT AND FELT
4- DECIDE WHAT YOU LIKED BEST OR FOUND THE MOST INTERESTING TO YOU
5- TELL ME ABOUT IT.

WHEN YOU ENTER THE MUSEUM, ASK FOR A FLOOR PLAN AT THE INFORMATION DESK. YOU ARE TO VISIT ALL OF THE ROOMS LISTED BELOW.

1ST FLOOR	2ND FLOOR
ROOM 7 - SMALL MAMMALS	ROOM 9 - ASIATIC MAMMALS
ROOM 9 - INVERTEBRATES	ROOM 13 - AFRICAN MAMMALS
ROOM 13 - OCEAN LIFE AND FISHES	
ROOM 19 - BIRDS	

3RD FLOOR	4TH FLOOR
ROOM 2 - PRIMATES	ROOM 9 - LATE DINOSAURS
ROOM 9 - REPTILES AND AMPHIBIANS	ROOM 13 - EARLY DINOSAURS

IN THE PAST STUDENTS HAVE GONE TO THE MUSEUM WITH OTHER MEMBERS OF THE CLASS BY CAR POOL OR BY BUS.

YOU ARE TO WRITE ABOUT THE SIX (6) SEPARATE EXHIBITS YOU FOUND TO BE THE MOST INTERESTING OR NOTEWORTHY TO YOU. EACH EXHIBIT MUST BE FROM A SEPARATE ROOM SELECTED FROM THOSE LISTED ABOVE. TELL ME ABOUT THE EXHIBIT AND WHY YOU LIKED IT OR FOUND IT INTERESTING. YOUR REPORT SHOULD NOT EXCEED 4 TYPED DOUBLE SPACED PAGES (@ 1,000 WORDS). YOU MUST ATTACH YOUR RECEIPT FOR ADMISSION TO THE FRONT CORNER OF YOUR REPORT.

ASSESSMENT:

YOUR REPORT IS DUE _____.

THE REPORT IS WORTH 30 POINTS AND WILL ASSESSED ON THE
COMPLETENESS AND QUALITY OF YOUR WORK.

THE IMPLEMENTATION:

THIS ASSIGNMENT IS HANDED OUT AND DISCUSSED AT THE FIRST LAB
SESSION OF THE SEMESTER. I REMIND THE STUDENTS ABOUT IT
PERIODICALLY DURING THE SEMESTER.

ASSIGNMENT # 20

GENERAL BIOLOGY II

LABORATORY EXERCISE #20 PROTISTA: GRAM STAINING OF BACTERIA

THE ASSIGNMENT:

THE PURPOSE OF THIS ASSIGNMENT IS FOR YOU TO FULLY UNDERSTAND THE GRAM STAINING PROCEDURE AND WHAT THE RESULTS INDICATE ABOUT THE STRUCTURE OF BACTERIA. YOU WILL BE PRACTICING YOUR ANALYTICAL AND COMMUNICATIVE ABILITIES AS WELL AS USING YOUR IMAGINATION.

BE AS IMAGINATIVE OR FANCIFUL AS YOU LIKE IN YOUR NARRATIVE.

YOU ARE TO PRETEND THAT YOU ARE A BACTERIUM AND WRITE THE GRAM STAINING PROCEDURE FROM THE BACTERIUM'S POINT OF VIEW. YOU SHOULD ASK YOURSELF THE FOLLOWING QUESTIONS:

1- WHAT HAPPENED TO YOU AT EACH STEP OF THE PROCEDURE AND WHY?
2- WHAT COLOR ARE YOU AT THE END OF THE PROCESS?
3- WHAT DOES THIS TELL YOU ABOUT YOUR STRUCTURE?

THIS ASSIGNMENT SHOULD BE 1 TO 2 PAGES (250-500 WORDS) IN LENGTH, EITHER TYPED OR WRITTEN CLEARLY.

ASSESSMENT:

THE ASSIGNMENT IS DUE AT THE NEXT LAB SESSION AND IS WORTH TEN POINTS.
THE ASSIGNMENT WILL BE ASSESSED ON THE COMPLETENESS OF INFORMATION AND QUALITY OF THE WORK.

THE IMPLEMENTATION:

THE ASSIGNMENT IS HANDED OUT AT THE BEGINNING OF THE LABORATORY SESSION AND DISCUSSED.
AT THE NEXT LAB SESSION EACH STUDENT READS THEIR WORK TO THEIR LABORATORY TEAM. THE TEAM DISCUSSES EACH WORK AND THEIR COMMENTS ARE RECORDED ON THE BACK OF EACH PAPER. EACH TEAM CHOOSES ONE ASSIGNMENT TO READ TO THE CLASS. THE TEAM PRESENTATIONS ARE FOLLOWED BY CLASS DISCUSSION AND ALL THE PAPERS ARE HANDED IN FOR MY ASSESSMENT.

ASSIGNMENT # 21

GENERAL BIOLOGY

LABORATORY EXERCISE #19 MONERA - LIFE CYCLE OF <u>PLASMODIUM</u>
<u>VIVAX</u>

THE ASSIGNMENT:

ONE WAY TO LEARN SOMETHING IS TO TEACH IT TO SOMEONE ELSE. IF YOU CAN TEACH SOMETHING TO ANOTHER PERSON SO THAT THE INFORMATION IS EASILY AND CLEARLY UNDERSTOOD THEN YOU KNOW THE INFORMATION YOURSELF.

THE PURPOSE OF THIS ASSIGNMENT IS FOR YOU TO BE ABLE TO UNDERSTAND THE LIFE CYCLE OF <u>PLASMODIUM</u>. THE CYCLE IS VERY COMPLEX AND THIS EXERCISE IS TO HELP YOU TO SEE THE OVERALL PATTERN. ONCE YOU HAVE ANALYZED THE CYCLE AND SEE THE PATTERN, THE DETAILS (I.E. SPECIFIC NAMES) WILL BE EASIER TO LEARN.

LAST SUNDAY YOUR FAVORITE TWIN COUSINS CAME OVER TO VISIT YOU BECAUSE THEY ARE HAVING TROUBLE IN THEIR JUNIOR HIGH SCHOOL BIOLOGY CLASS AND WANT YOUR HELP. THEIR TEACHER SPENT THE LAST CLASS DISCUSSING THE LIFE CYCLE OF <u>PLASMODIUM</u> AND HOW IT RELATES TO MAN (MALARIA) AND MOSQUITOES. YOUR COUSINS ARE VERY CONFUSED. EXPLAIN TO THEM, SO THEY CLEARLY UNDERSTAND, THE LIFE CYCLE OF <u>PLASMODIUM</u>, WHAT GOES ON INSIDE A HUMAN BEING WHO HAS MALARIA, AND WHAT CAUSES THE SYMPTOMS OF THE DISEASE AND WHERE THE MOSQUITO FITS INTO THE PICTURE. REMEMBER YOUR COUSINS ARE ONLY 14 YEARS OLD AND NOT COLLEGE STUDENTS.

THIS ASSIGNMENT SHOULD BE NO MORE THAN 1 OR 2 PAGES (@ 250-500 WORDS) IN LENGTH.

ASSESSMENT:

THE ASSIGNMENT IS DUE AT THE NEXT LAB SESSION AND IS WORTH TEN POINTS.
THE ASSIGNMENT WILL BE ASSESSED ON COMPLETENESS AND CLARITY.

THE IMPLEMENTATION:

THE ASSIGNMENT IS HANDED OUT AT THE END OF THE LAB SESSION AND DISCUSSED. AT THE NEXT LAB SESSION PARTNERS EXCHANGE AND READ EACH OTHERS WORK AND RECORD THEIR COMMENTS. ALL PAPERS ARE HANDED IN FOR MY ASSESSMENT.

ASSIGNMENT # 22

GENERAL BIOLOGY II

LABORATORY EXERCISE #30   FUNGI - RHIZOPUS LIFE CYCLE

THE ASSIGNMENT:

THE PURPOSE OF THIS ASSIGNMENT IS TO:

1- GIVE YOU PRACTICE IN OBSERVATION, ANALYSIS AND SYNTHESIS
2- HAVE YOU UNDERSTAND THE LIFE CYCLE OF RHIZOPUS
3- GIVE YOU AN UNDERSTANDING OF THE RELATIONSHIP BETWEEN MITOSIS AND MEIOSIS IN THE ALTERNATION OF GENERATION OF RHIZOPUS.

THE ASSIGNMENTS HAS THREE PARTS.

PART I

FROM THE SLIDES AVAILABLE IN CLASS, YOU ARE TO MAKE FULLY LABELED SKETCHS OF THE VARIOUS STAGES OF THE RHIZOPUS LIFE CYCLE. YOU ARE TO INDICATE ON THE SKETCH IF A STRUCTURE IS DIPLOID (2N) OR HAPLOID (N).

PART II

MAKE A LABELED DIAGRAM OF THE LIFE CYCLE OF RHIZOPUS. INDICATE WHICH STAGES OF THE LIFE CYCLE ARE SEXUAL AND ASEXUAL.

PART III

EXPLAIN TO YOUR BEST FRIEND HOW THE LIFE CYCLE WORKS IN TERMS OF SEXUAL AND ASEXUAL STAGES AND THE ADVANTAGES AND DISADVANTAGES OF EACH TYPE OF REPRODUCTION. THIS EXPLANATION SHOULD BE FROM 1/2 TO 1 PAGE (150-250 WORDS) IN LENGTH.

ASSESSMENT:

ALL THREE PARTS OF THE ASSIGNMENT ARE DUE AT THE NEXT LAB SESSION.
THE WHOLE ASSIGNMENT IS WORTH 15 POINTS.
THE ASSIGNMENT WILL BE ASSESSED ON COMPLETENESS AND QUALITY OF WORK.

THE IMPLEMENTATION:

THE ASSIGNMENT IS HANDED OUT AT THE BEGINNING OF THE LAB SESSION AND DISCUSSED.

ASSIGNMENT # 23

GENERAL BIOLOGY II

LABORATORY EXERCISE #35 Seed Vascular Plants

THE ASSIGNMENT:

The purpose of this assignment is to increase your:

1- observational skills
2- ability to record what you observe
3- ability to describe, associate and compare what you have observed.

This assignment has four parts. All three drawings are to follow the drawing format handed out the first laboratory session (see Appendix I, assignment #4).

PART I

Make a detailed drawing of a vascular bundle from the dicot stem slide. On the back of the paper sketch the whole dicot stem in cross section along with the usual information.

PART II

Make a detailed drawing of a vascular bundle from the monocot stem slide. On the back of the paper sketch the whole monocot stem in cross section along with the usual information.

PART III

For the comparison portion of your work compare the dicot vascular bundle to the monocot vascular bundle.

PART IV

Draw a section of a dicot leaf in cross section. Be sure that the section you select includes a vascular bundle. Compare the structure of the leaf to the structure of the dicot stem.

ASSESSMENT:

All four parts of the assignment are DUE at the next laboratory session.
The assignment is worth 15 points and will be assessed on completeness and clarity.

THE IMPLEMENTATION:

The assignment is handed out at the beginning of the lab

SESSION AND DISCUSSED.

ASSIGNMENT # 24

GENERAL BIOLOGY II

LABORATORY EXERCISE #8 ANIMAL TISSUE WITH HANDOUT

THE ASSIGNMENT:

THE PURPOSE OF THIS ASSIGNMENT IS TO HELP YOU INCREASE YOUR:

1- OBSERVATIONAL SKILLS
2- RECORDING SKILLS
3- ABILITY TO DESCRIBE AND COMPARE WHAT YOU HAVE OBSERVED.

YOU ARE TO MAKE SKETCHES OF ALL THE ANIMAL TISSUES AVAILABLE ACCORDING TO EXERCISE #8 AND THE DETAILED HANDOUT ON ANIMAL TISSUES. FOR EACH TISSUE YOU ARE TO SKETCH AN OVERVIEW OF THE TISSUE AT LOW POWER AND A DETAILED VIEW OF A FEW CELLS AT HIGH POWER. FOR EACH TISSUE DESCRIBE, ASSOCIATE AND COMPARE AS USUAL. COMPARE THE TISSUES OF THE SAME GROUP (E.G. COMPARE ALL THREE TYPES OF EPITHELIUM).

ASSESSMENT:

THIS ASSIGNMENT IS DUE IN TWO WEEKS AND IS WORTH 30 POINTS. THIS ASSIGNMENT WILL BE ASSESSED ON COMPLETENESS AND CLARITY.

THE IMPLEMENTATION:

THE ASSIGNMENT IS HANDED OUT AT THE BEGINNING OF THE LAB SESSION AND DISCUSSED WITH THE STUDENTS.

ASSIGNMENT # 25

GENERAL BIOLOGY II

LABORATORY EXERCISE #15 FLOWERING PLANT DEVELOPMENT

THE ASSIGNMENT:

THE PURPOSE OF THIS ASSIGNMENT IS TO:

1- INCREASE YOUR OBSERVATIONAL SKILLS
2- GIVE YOU PRACTICE IN ANALYZING INFORMATION AND MAKING COMPARISONS.

THE INFORMATION NECESSARY TO DO THIS ASSIGNMENT IS ON PAGE 107 IN THE LAB MANUAL AND IN YOUR TEXT BOOK.

YOU ARE HAVING A DISCUSSION ABOUT PLANT GERMINATION WITH YOUR FRIEND WHO THINKS THAT ALL PLANTS GERMINATE THE SAME WAY. EXPLAIN AND COMPARE THE DIFFERENT TYPES OF GERMINATION TO YOUR FRIEND. BE SURE TO INCLUDE IN YOUR EXPLANATION A) WHERE THE COTYLEDONS END UP IN RELATIONSHIP TO THE SOIL LEVEL, B) WHICH PART OF THE SEEDLING ELONGATES FIRST, AND C) HOW THE SHOOT IS PROTECTED. THE TERMINOLOGY WILL BE NEW TO YOUR FRIEND SO BE SURE TO DEFINE EACH TERM.

THIS ASSIGNMENT SHOULD BE ABOUT 2 PAGES (@ 500 WORDS) IN LENGTH.

THE ROUGH DRAFT OF THE ASSIGNMENT IS DUE AT THE BEGINNING OF THE NEXT LAB SESSION WHEN YOU WILL READ YOUR WORK TO YOUR TEAM MEMBERS. TEAM COMMENTS WILL BE RECORDED ON THE BACK OF THE DRAFT. TEAM MEMBERS SHOULD CHECK FOR CORRECTNESS AND COMPLETENESS OF INFORMATION AND UNDERSTANDABILITY.

ASSESSMENT:

THE FINAL ASSIGNMENT IS DUE THE FOLLOWING LAB SESSION ALONG WITH THE ROUGH DRAFT.
THE ASSIGNMENT IS WORTH TEN POINTS.
THE ASSIGNMENT WILL BE ASSESSED ON COMPLETENESS AND QUALITY OF WORK.

THE IMPLEMENTATION:

THE ASSIGNMENT IS HANDED OUT IN CLASS AND DISCUSSED.
AT THE BEGINNING OF THE NEXT LAB SESSION TEAM MEMBERS READ THEIR WORK TO EACH OTHER, DISCUSS THE WORK AND RECORD THEIR COMMENTS ON THE BACK OF THE ASSIGNMENTS.
THE FOLLOWING WEEK THE ROUGH DRAFTS AND FINAL ASSIGNMENTS ARE HANDED IN FOR ASSESSMENT.

ASSIGNMENT # 26

GENERAL BIOLOGY II

LABORATORY EXERCISE   ECOLOGY GAME

THE ASSIGNMENT:

THE PURPOSE OF THIS ASSIGNMENT IS TO GIVE YOU EXPERIENCE IN SYNTHESIZING INFORMATION.

YOU HAVE GONE HOME TO VISIT YOUR FAMILY FOR THE WEEKEND AND OVER DINNER THEY ASK YOU WHAT YOU DID IN BIOLOGY LAB THIS WEEK, YOU REPLY THAT YOU PLAYED THE ECOLOGY GAME. THERE ARE VARIOUS COMMENTS MADE ABOUT THE LAB BEING A WASTE OF YOUR TIME AND MONEY. EXPLAIN TO YOUR FAMILY WHICH PRINCIPLES OF ECOLOGY AND EVOLUTION WERE LEARNED OR DEMONSTRATED BY PLAYING THE GAME. USE SPECIFIC EXAMPLES FROM THE GAME YOU PLAYED TO MAKE YOUR POINTS CLEAR.

THIS ASSIGNMENT SHOULD BE ABOUT 1-2 PAGES (250-500 WORDS) IN LENGTH.

ASSESSMENT:

THE ASSIGNMENT IS DUE AT THE NEXT LAB SESSION AND IS WORTH TEN POINTS.
THE ASSIGNMENT WILL BE ASSESSED ON COMPLETENESS AND QUALITY OF YOUR WORK.

THE IMPLEMENTATION:

THE ASSIGNMENT IS HANDED OUT AND DISCUSSED PRIOR TO PLAYING THE GAME.

ASSIGNMENT # 27

GENERAL BIOLOGY II

LABORATORY EXERCISE   WATERFALL FIELD TRIP

THE ASSIGNMENT:

THIS ASSIGNMENT HAS SEVERAL PURPOSES:

1- TO INCREASE YOUR ABILITY TO OBSERVE "SEE" THE ENVIRONMENT
2- TO GIVE YOU AN OPPORTUNITY TO RECORD WHAT YOU OBSERVE, HEAR AND FEEL
3- TO INCREASE YOU AWARENESS OF THE COMPLEXITY OF THE ENVIRONMENT AND YOUR REACTIONS TO IT
4- TO GIVE YOU THE OPPORTUNITY TO REFLECT ON THE ENTIRE EXPERIENCE.

THIS ASSIGNMENT HAS TWO PARTS.

PART I JOURNAL

YOU ARE TO CHOOSE A SPECIFIC OUTDOOR ENVIRONMENT (YOUR SPOT) AND VISIT IT DAILY FOR TWO WEEKS. EACH VISIT SHOULD BE FOR AT LEAST 30 MINUTES. I WANT YOU TO OBSERVE AND HEAR WHAT IS THERE AND TO THINK ABOUT (REFLECT ON) WHAT YOU ARE OBSERVING AND FEELING. TO HELP YOU "BE THERE" MENTALLY YOU ARE TO KEEP A JOURNAL ABOUT WHAT YOU "SEE", HEAR AND FEEL DURING YOUR VISITS. AS YOU GO TO YOUR SPOT FOR TWO WEEKS YOU WILL BEGIN TO NOTICE SUBTLE CHANGES IN THE PLANTS, ANIMALS, SOIL, AND WEATHER. THE OBSERVATIONS THAT YOU MAKE AND RECORD WILL COMPRISE YOUR JOURNAL.

PART II LETTER

AT THE END OF TWO WEEKS YOU ARE TO REVIEW YOUR JOURNAL, REFLECT ON YOUR EXPERIENCE AND THEN WRITE A LETTER TO A FRIEND TELLING THEM ABOUT YOUR EXPERIENCE AND WHAT IT HAS MEANT TO YOU. YOUR LETTER SHOULD INCLUDE: YOUR THOUGHTS AND FEELINGS AND HOW THEY CHANGED OVER TIME, WHETHER YOU WOULD RECOMMEND THE EXPERIENCE TO YOUR FRIEND AND WHY. THE LETTER SHOULD BE AT LEAST TWO PAGES (500 WORDS) IN LENGTH.

TO HELP YOU ANALYZE YOUR EXPERIENCE I HAVE INCLUDED A COUPLE OF QUESTIONS FOR YOU TO THINK ABOUT:

1- HOW DID YOU FEEL ABOUT THE ASSIGNMENT AT THE BEGINNING OF THE TWO WEEKS?
2- HOW DO YOU FEEL ABOUT THE ASSIGNMENT NOW THAT IT IS OVER?
3- WHY DID OR DIDN'T YOU ENJOY THE ASSIGNMENT?
4- HOW DID YOU FEEL WHEN YOU WERE AT YOUR SPOT?
5- WHAT WAS THE ONE MAJOR CHANGE YOU OBSERVED IN YOUR SPOT OVER THE TWO WEEK PERIOD?

6- HOW COULD THE ASSIGNMENT BE IMPROVED AS FAR AS YOUR ARE
   CONCERNED?

ASSESSMENT:

PARTS I AND II ARE DUE IN THREE WEEKS AND ARE WORTH 30
POINTS.
   YOUR WORKS WILL BE ASSESSED ON COMPLETENESS OF TASK AND THE
SERIOUSNESS OF YOUR EFFORT.

THE IMPLEMENTATION:

THE ASSIGNMENT IS HANDED OUT AND DISCUSSED.

ASSIGNMENT # 28

GENERAL BIOLOGY II

LABORATORY EXERCISE   FETAL PIG DISSECTION: I

THE ASSIGNMENT:

THE PURPOSE OF THIS ASSIGNMENT IS TO FOCUS YOUR OBSERVATIONS AND THOUGHTS WHILE YOU ARE DISSECTING YOUR FETAL PIG.

TO DO THE DISSECTION YOU AND YOUR PARTNER WILL NEED DISSECTING EQUIPMENT, A DISSECTING MICROSCOPE, A LIGHT MICROSCOPE, A DISSECTION MANUAL AND A FETAL PIG.

IT IS IMPORTANT FOR YOU TO OBSERVE AND UNDERSTAND:

1- THE STRUCTURE AND FUNCTION OF EACH ORGAN
2- THE STRUCTURAL AND FUNCTIONAL RELATIONSHIPS BETWEEN THE ORGAN AND ITS ORGAN SYSTEM
3- THE STRUCTURAL AND FUNCTIONAL RELATIONSHIP OF EACH ORGAN SYSTEM TO THE ORGANISM.

YOU WILL USE THE FOLLOWING FORMAT TO STUDY THE PIG AS YOU DO THE DISSECTION. ALTHOUGH THE ASSIGNMENT IS FOR ONLY THREE ORGANS, YOU SHOULD BE ABLE TO ANSWER THE SEVEN PARTS FOR ALL THE ORGANS AND SYSTEMS DISSECTED. ANSWER PARTS 1-7 IN COMPLETE SENTENCES USING YOUR OWN WORDS.

PARTS

1- DESCRIBE FOR A HIGH SCHOOL SENIOR HOW TO FIND OR LOCATE EACH PART OR ORGAN. IN YOUR DESCRIPTION YOU SHOULD USE THE CORRECT TERMINOLOGY (E.G. POSTERIOR, CAUDAL, ANTERIOR, CRANIAL, VENTRAL, DORSAL, LATERAL) TO PLACE THE ORGAN RELATIVE TO ADJACENT ORGANS AND WITHIN THE WHOLE ORGANISM.

2- IN WHICH BODY CAVITY DOES THE ORGAN OR PART LIE?

3- MAKE A FULLY LABELED SKETCH OF THE ORGAN OR PART. DESCRIBE HOW IT LOOKS (E.G. SIZE, SHAPE, COLOR, TEXTURE).

4- ASSOCIATE THE ORGAN OR PART TO SOMETHING ELSE IN YOUR EXPERIENCE.

5- HOW DOES THE STRUCTURE OF THE ORGAN OR PART RELATE TO ITS FUNCTION?

6- TO WHAT ORGAN SYSTEM DOES THE ORGAN OR PART BELONG?

7- HOW DOES THE ORGAN OR PART FIT INTO THE ORGAN SYSTEM?

ANSWER PARTS 1-7 FOR:

1- LEFT LUNG
2- SPLEEN
3- GALL BLADDER

ASSESSMENT:

THE ASSIGNMENT IS DUE AT THE NEXT LAB SESSION AND IS WORTH 12 POINTS.
YOUR WORK WILL BE ASSESSED BASED ON THE COMPLETENESS AND CORRECTNESS OF YOUR WORK.

THE IMPLEMENTATION:

THE ASSIGNMENT IS HANDED OUT AND DISCUSSED AT THE BEGINNING OF THE LAB SESSION.

ASSIGNMENT # 29

GENERAL BIOLOGY II

LABORATORY EXERCISE   FETAL PIG DISSECTION: II

THE ASSIGNMENT:

THE PURPOSE OF THIS ASSIGNMENT IS TO FOCUS YOUR OBSERVATIONS AND THOUGHTS WHILE YOU ARE DISSECTING YOUR FETAL PIG.

TO DO THE DISSECTION YOU AND YOUR PARTNER WILL NEED DISSECTING EQUIPMENT, A DISSECTING MICROSCOPE, A LIGHT MICROSCOPE, A DISSECTION MANUAL AND A FETAL PIG.

IT IS IMPORTANT FOR YOU TO OBSERVE AND UNDERSTAND:

1- THE STRUCTURE AND FUNCTION OF EACH ORGAN
2- THE STRUCTURAL AND FUNCTIONAL RELATIONSHIPS BETWEEN THE ORGAN AND ITS ORGAN SYSTEM
3- THE STRUCTURAL AND FUNCTIONAL RELATIONSHIP OF EACH ORGAN SYSTEM TO THE ORGANISM.

YOU WILL USE THE FOLLOWING FORMAT TO STUDY THE PIG AS YOU DO THE DISSECTION. ALTHOUGH THE ASSIGNMENT IS FOR ONLY THREE ORGANS, YOU SHOULD BE ABLE TO ANSWER THE SEVEN PARTS FOR ALL THE ORGANS AND SYSTEMS DISSECTED. ANSWER PARTS 1-7 IN COMPLETE SENTENCES USING YOUR OWN WORDS.

PARTS

1- DESCRIBE FOR A HIGH SCHOOL SENIOR HOW TO FIND OR LOCATE EACH PART OR ORGAN. IN YOUR DESCRIPTION YOU SHOULD USE THE CORRECT TERMINOLOGY (E.G. POSTERIOR, CAUDAL, ANTERIOR, CRANIAL, VENTRAL, DORSAL, LATERAL) TO PLACE THE ORGAN RELATIVE TO ADJACENT ORGANS AND WITHIN THE WHOLE ORGANISM.

2- IN WHICH BODY CAVITY DOES THE ORGAN OR PART LIE?

3- MAKE A FULLY LABELED SKETCH OF THE ORGAN OR PART. DESCRIBE HOW IT LOOKS (E.G. SIZE, SHAPE, COLOR, TEXTURE).

4- ASSOCIATE THE ORGAN OR PART TO SOMETHING ELSE IN YOUR EXPERIENCE.

5- HOW DOES THE STRUCTURE OF THE ORGAN OR PART RELATE TO ITS FUNCTION?

6- TO WHAT ORGAN SYSTEM DOES THE ORGAN OR PART BELONG?

7- HOW DOES THE ORGAN OR PART FIT INTO THE ORGAN SYSTEM?

Answer PARTS 1-7 for:

1- right common carotid artery
2- left renal vein
3- left atrium

ASSESSMENT:

The assignment is DUE at the next lab session and is worth 12 points.
Your work will be assessed based on the completeness and correctness of your work.

THE IMPLEMENTATION:

The assignment is handed out and discussed at the beginning of the lab session.

ASSIGNMENT # 30

GENERAL BIOLOGY II

LABORATORY EXERCISE   FETAL PIG DISSECTION: III

THE ASSIGNMENT:

THE PURPOSE OF THIS ASSIGNMENT IS TO FOCUS YOUR OBSERVATIONS AND THOUGHTS WHILE YOU ARE DISSECTING YOUR FETAL PIG.

TO DO THE DISSECTION YOU AND YOUR PARTNER WILL NEED DISSECTING EQUIPMENT, A DISSECTING MICROSCOPE, A LIGHT MICROSCOPE, A DISSECTION MANUAL AND A FETAL PIG.

IT IS IMPORTANT FOR YOU TO OBSERVE AND UNDERSTAND:

1- THE STRUCTURE AND FUNCTION OF EACH ORGAN
2- THE STRUCTURAL AND FUNCTIONAL RELATIONSHIPS BETWEEN THE ORGAN AND ITS ORGAN SYSTEM
3- THE STRUCTURAL AND FUNCTIONAL RELATIONSHIP OF EACH ORGAN SYSTEM TO THE ORGANISM.

YOU WILL USE THE FOLLOWING FORMAT TO STUDY THE PIG AS YOU DO THE DISSECTION. ALTHOUGH THE ASSIGNMENT IS FOR ONLY THREE ORGANS, YOU SHOULD BE ABLE TO ANSWER THE SEVEN PARTS FOR ALL THE ORGANS AND SYSTEMS DISSECTED. ANSWER PARTS 1-7 IN COMPLETE SENTENCES USING YOUR OWN WORDS.

PARTS

1- DESCRIBE FOR A HIGH SCHOOL SENIOR HOW TO FIND OR LOCATE EACH PART OR ORGAN. IN YOUR DESCRIPTION YOU SHOULD USE THE CORRECT TERMINOLOGY (E.G. POSTERIOR, CAUDAL, ANTERIOR, CRANIAL, VENTRAL, DORSAL, LATERAL) TO PLACE THE ORGAN RELATIVE TO ADJACENT ORGANS AND WITHIN THE WHOLE ORGANISM.

2- IN WHICH BODY CAVITY DOES THE ORGAN OR PART LIE?

3- MAKE A FULLY LABELED SKETCH OF THE ORGAN OR PART. DESCRIBE HOW IT LOOKS (E.G. SIZE, SHAPE, COLOR, TEXTURE).

4- ASSOCIATE THE ORGAN OR PART TO SOMETHING ELSE IN YOUR EXPERIENCE.

5- HOW DOES THE STRUCTURE OF THE ORGAN OR PART RELATE TO ITS FUNCTION?

6- TO WHAT ORGAN SYSTEM DOES THE ORGAN OR PART BELONG?

7- HOW DOES THE ORGAN OR PART FIT INTO THE ORGAN SYSTEM?

ANSWER PARTS 1-7 FOR:

1- RIGHT KIDNEY
2- URINARY BLADDER
3- LEFT OVARY OR TESTIS

ASSESSMENT:

THE ASSIGNMENT IS DUE AT THE NEXT LAB SESSION AND IS WORTH 12 POINTS.
YOUR WORK WILL BE ASSESSED BASED ON THE COMPLETENESS AND CORRECTNESS OF YOUR WORK.

THE IMPLEMENTATION:

THE ASSIGNMENT IS HANDED OUT AND DISCUSSED AT THE BEGINNING OF THE LAB SESSION.

APPENDIX III

WRITING ASSIGNMENTS FOR GENERAL BIOLOGY LECTURE (#31-95)

ASSIGNMENT # 31

GENERAL BIOLOGY I AND II

LECTURE ANCHORING

TOPIC   INTRODUCTION

THE ASSIGNMENT:

ASK THE CLASS TO THINK ABOUT THE FOLLOWING QUESTIONS: HOW DO
YOU LEARN? HOW DO YOU STUDY FOR AN EXAM? WHAT CLASSES DID YOU
ENJOY THE MOST IN HIGH SCHOOL? WHY DID YOU ENJOY IT OR THEM? IN
WHICH CLASS(ES) DID YOU DO YOUR BEST? WHAT DID THE TEACHER DO
THAT YOU LIKED?

THE IMPLEMENTATION:

AFTER THEY HAVE THOUGHT ABOUT THE QUESTIONS THERE IS AN
INDIVIDUAL 5 MINUTE FREE WRITING EXERCISE. SINCE IT THE FIRST
TIME THIS SEMESTER THE CLASS HAS DONE FREE WRITING I EXPLAIN WHAT
IT IS, HOW IT IS DONE, AND THAT THERE ARE NO RIGHT ANSWERS OR
GRADES. AFTER FIVE MINUTES THE STUDENT EXCHANGES PAPERS WITH
THEIR NEIGHBOR AND READS THEIR FREE WRITING. BY EXCHANGING
PAPERS THE STUDENT BEGINS TO UNDERSTAND THAT NOT EVERYONE LEARNS
THE SAME WAY. I ASK FOR VOLUNTEERS TO READ THEIR PAPERS TO THE
CLASS AND THIS LEADS INTO THE DISCUSSION ON LEARNING STYLES.
AFTER THE DISCUSSION I HELP THE THE CLASS FORM INTO STUDY GROUPS.
I THEN COLLECT THE STUDY GROUPS' LISTS OF NAMES AND PHONE NUMBER
AND THE FREE WRITING. I READ AND MAKE SUPPORTIVE COMMENTS ON THE
FREE WRITINGS AND RETURN THEM AT THE BEGINNING OF THE NEXT
LECTURE SESSION.

125

ASSIGNMENT # 32

GENERAL BIOLOGY I

LECTURE ANCHORING

TOPIC   DEFINITION OF LIFE

THE ASSIGNMENT:

Ask each peer group (4 or 5 students) to make up a list of characteristics or attributes that are common to all living things (plants and animals).

THE IMPLEMENTATION:

Each group works on their list for 3-5 minutes. I circulate as they work to help get them started and answer any questions. After they have finished making their lists, I have each group read theirs to the class and as the discussion progresses I make a master list on the black board. I rephrase or group attributes to end up with a complete list. This leads directly into the lecture. I do not collect the lists.

ASSIGNMENT # 33

GENERAL BIOLOGY I

LECTURE ANCHORING

TOPIC   SCIENTIFIC METHOD

THE ASSIGNMENT:

ASK THE CLASS TO THINK ABOUT THE FOLLOWING QUESTION: HOW DO
YOU SOLVE A PROBLEM? (E.G. HOW DO YOU BUY A CAR? HOW DID YOU
CHOOSE THIS COLLEGE? HOW DID YOU DECIDE WHAT TO HAVE FOR
BREAKFAST OR WHAT TO WEAR TODAY?) CHOOSE A PROBLEM AND WRITE
DOWN STEP BY STEP HOW YOU SOLVED IT.

THE IMPLEMENTATION:

THIS IS AN INDIVIDUAL FREE WRITING EXERCISE FOR 3-5 MINUTES.
AFTER THE TIME IS UP I ASK FOR 3 OR 4 VOLUNTEERS TO READ THEIR
WORKS TO THE CLASS. THE CLASS THEN DISCUSSES WHAT SIMILAR STEPS
OR ACTIONS ALL OF THESE STUDENTS TOOK TO SOLVE THEIR PROBLEMS. I
PUT THESE COMMON STEPS ON THE BLACK BOARD AND REFER TO THEM AS I
LECTURE ON THE SCIENTIFIC METHOD AND PROBLEM SOLVING.
I DO NOT COLLECT THIS EXERCISE AS I AM GOING TO GIVE THEM A
HOMEWORK ASSIGNMENT ON THE SCIENTIFIC METHOD AND PROBLEM SOLVING
(SEE ASSIGNMENT #34).

ASSIGNMENT # 34

GENERAL BIOLOGY I

LECTURE SUMMATION

TOPIC   SCIENTIFIC METHOD

THE ASSIGNMENT:

FOR THIS HOMEWORK ASSIGNMENT YOU ARE TO CHOOSE AN EXAMPLE OF
YOUR OWN (REAL OR NOT) AND WRITE THE:

1- OBSERVATIONS (DATA)

2- INDUCTIVE GENERALIZATION

3- GENERALIZATION HYPOTHESIS

4- CAUSAL EXPLANATORY HYPOTHESIS.

THIS ASSIGNMENT IS DUE AT THE BEGINNING OF THE NEXT LECTURE
SESSION.

I WILL COLLECT THIS ASSIGNMENT AND MAKE COMMENTS. A CHECK ON
THE TOP OF YOUR PAPER MEANS THAT EVERYTHING IS OK. IF YOU DO NOT
RECEIVE A CHECK THE ASSIGNMENT MUST BE REPEATED WITH A NEW
EXAMPLE. THE NEW EXAMPLE ALONG WITH THE PREVIOUS ASSIGNMENT IS
HANDED IN AT THE NEXT LECTURE SESSION.

THE IMPLEMENTATION:

THIS ASSIGNMENT IS HANDED OUT AFTER THE LECTURE AND
DISCUSSION ON THE SCIENTIFIC METHOD. AT THE BEGINNING OF THE
NEXT LECTURE SESSION EACH STUDENT READS THEIR EXAMPLE TO A PEER
GROUP OF 4-5 STUDENTS. THE GROUP COMMENTS ON WHETHER THE EXAMPLE
IS CLEAR, IF EACH PART IS CORRECT FOR THE EXAMPLE, AND GIVES
SUGGESTIONS ON HOW TO FIX ANY PROBLEMS. ALL COMMENTS AND
SUGGESTIONS ARE RECORDED ON THE BACK OF THE HOMEWORK PAPER.
I DO AN EXAMPLE ON THE BLACK BOARD AND ANSWER QUESTIONS FROM
THE CLASS.
I COLLECT THESE ASSIGNMENTS, READ THEM OVER AND MAKE MY OWN
COMMENTS AND SUGGESTIONS. I HAND BACK THE ASSIGNMENT AT THE NEXT
LECTURE SESSION AND THE STUDENTS WHO HAVE HAD PROBLEMS ARE GIVEN
THE ASSIGNMENT TO REDO WITH A NEW EXAMPLE. STUDENTS REPEAT THE
ASSIGNMENT UNTIL THEY RECEIVE A CHECK. WHEN A STUDENT REPEATS
THE ASSIGNMENT THEY HAND IN ALL THE PREVIOUS ATTEMPTS WITH THEIR
NEWEST EFFORT.
I PLACE A CHECK ON THE TOP OF THOSE ASSIGNMENTS WHICH DO NOT
HAVE TO BE REPEATED. I KEEP TRACK OF WHO RECEIVES A CHECK UNTIL
THE ENTIRE CLASS HAS BEEN SUCCESSFUL.

ASSIGNMENT # 35

GENERAL BIOLOGY I

LECTURE ANCHORING

TOPIC   ORIGIN OF LIFE

THE ASSIGNMENT:

CLUSTER ON THE "ORIGIN OF LIFE" OR HOW DID LIFE GET HERE?

THE IMPLEMENTATION:

SINCE THIS IS THE FIRST TIME I HAVE USED CLUSTERING IN CLASS THIS SEMESTER, I WILL FIRST EXPLAIN THE PROCESS AND GIVE AN EXAMPLE.

EACH STUDENT CLUSTERS FOR 3-5 MINUTES ON THE TOPIC. THE STUDENT THEN EXCHANGES CLUSTERS WITH THEIR NEIGHBOR AND UNDERLINES TWO IDEAS THAT WERE NOT THEIR CLUSTERS OR IDEAS THAT ARE CONNECTED IN DIFFERENT WAYS.

BEFORE I CONTINUE WITH THE CLASS DISCUSSION ON THE ORIGIN OF LIFE, I DISCUSS THE USE OF CLUSTERING AS ONE APPROACH TO ESSAY EXAM QUESTIONS. I SHOW THEM MY CLUSTER AS AN EXAMPLE.

I THEN ASK THE CLASS TO GIVE ME IDEAS, WORDS, CONCEPTS, PHRASES OR CONNECTIONS FROM THEIR CLUSTERS. I MAKE A LIST OF THESE ON THE BLACK BOARD AND THEN ORGANIZE THEM FOR THE LECTURE. I DO NOT COLLECT THIS EXERCISE.

ASSIGNMENT # 36

GENERAL BIOLOGY I

LECTURE ANCHORING

TOPIC   EVOLUTION

THE ASSIGNMENT:

CLUSTER ON THE WORD "EVOLUTION".

THE IMPLEMENTATION:

THIS IS A 3-5 MINUTE INDIVIDUAL CLUSTERING EXERCISE. AFTER THE CLUSTERING IS FINISHED PEER GROUPS (4-5 STUDENTS) ARE FORMED. PAPERS ARE EXCHANGED AND EACH STUDENT UNDERLINES ONE WORD, IDEA OR PHRASE ON EACH PAPER THAT IS DIFFERENT FROM THEIR OWN CLUSTER. I CIRCULATE THROUGH THE ROOM AS THE GROUPS WORK AND MAKE A COMMENT HERE AND THERE ON ANY NOVEL IDEAS OR CONNECTIONS. I THEN ASK FOR EACH GROUP TO GIVE ME WORDS, IDEAS OR PHRASES FROM THEIR PAPERS WHICH I PUT ON THE BLACK BOARD. AS I PUT THE WORDS ON THE BOARD I GROUP THEM WHERE POSSIBLE AND REFER TO THEM AS I LECTURE.

ASSIGNMENT # 37

GENERAL BIOLOGY I

LECTURE SUMMATION

TOPIC   EVOLUTION

THE ASSIGNMENT:

THE PURPOSE OF THIS HOMEWORK ASSIGNMENT IS TO GIVE YOU
PRACTICE EXPLAINING THE THEORY OF EVOLUTION IN YOUR OWN WORDS.

USING AN EXAMPLE OF YOUR OWN CHOICE, CLEARLY EXPLAIN HOW
DARWIN'S THEORY OF EVOLUTION WORKS. BE SURE TO USE THE FOLLOWING
TERMS: TIME, CHANCE, VARIATION, SELECTIVE PRESSURE, DIFFERENTIAL
REPRODUCTION, ADAPTATION, POPULATION, AND INDIVIDUAL. BE AS
CREATIVE AS YOU LIKE WITH YOUR EXAMPLE, YOU ARE NOT LIMITED TO
REALITY.

ASSESSMENT:

THE ASSIGNMENT SHOULD BE 1-2 PAGES (@250-500 WORDS) IN
LENGTH.
THIS ASSIGNMENT IS DUE AT THE NEXT LECTURE SESSION.
THIS ASSIGNMENT WILL BE ASSESSED ON THE CORRECT USAGE OF ALL
TERMS AND WHETHER YOUR EXPLANATION DEMONSTRATES A CLEAR
UNDERSTANDING OF HOW EVOLUTION WORKS. THE ASSIGNMENTS WILL BE
RETURNED WITH COMMENTS.

SCORING GUIDE:

3- DID THE ASSIGNMENT, CORRECT USAGE, WORK DEMONSTRATES CLEAR
   UNDERSTANDING
2- DID THE ASSIGNMENT, SOME ERRORS, WORK DEMONSTRATES SOME
   CONFUSION ABOUT PROCESS
1- DID THE ASSIGNMENT BUT WORK CURSORY AND CONFUSED
0- DID NOT DO THE ASSIGNMENT

THE IMPLEMENTATION:

THIS ASSIGNMENT WILL BE HANDED OUT AND DISCUSSED AT THE END
OF THE EVOLUTION LECTURE.

ASSIGNMENT # 38

GENERAL BIOLOGY I

LECTURE HOMEWORK

TOPIC   EISELEY-DARWIN

THE ASSIGNMENT:

THE PURPOSE OF THIS HOMEWORK ASSIGNMENT IS FOR YOU TO HAVE
PRACTICE IN ANALYZING AND SYNTHESIZING INFORMATION ABOUT SCIENCE
AS SEEN FROM AN ANTHROPOLOGIST'S VIEW POINT.

YOU ARE TO READ CHAPTER 18 "THE DANCERS IN THE RING" IN LOREN
EISELEY'S ALL THE STRANGE HOURS AND THEN IDENTIFY AND DISCUSS THE
MAJOR POINT OR IDEA THAT LOREN EISELEY IS MAKING ABOUT HOW
SCIENCE WORKS. GIVE TWO SPECIFIC EXAMPLES FROM THE CHAPTER OF
HOW EISELEY USES DARWIN TO SUPPORT HIS MAJOR POINT.

THE AUDIENCE FOR THIS ASSIGNMENT IS YOUR BEST FRIEND. YOU
WANT TO WRITE YOUR ASSIGNMENT SO THAT THEY WILL UNDERSTAND.

THIS ASSIGNMENT IS DUE IN TWO WEEKS.

THIS ASSIGNMENT SHOULD BE 2-3 PAGES (500-750 WORDS) IN
LENGTH.

ASSESSMENT:

SCORING GUIDE:

3- DID THE TASK, SHOWS SERIOUS EFFORT
2- DID THE TASK, WORK HASTY, BRIEF, SHOWS LITTLE THOUGHT
1- CURSORY WORK
0- WORK NOT DONE

THE IMPLEMENTATION:

THE ASSIGNMENT IS HANDED OUT IN CLASS AND DISCUSSED. THE DAY
THE PAPERS ARE DUE THE CLASS WILL WORK IN PEER GROUPS. EACH
STUDENT READS THEIR ASSIGNMENT TO THE GROUP AND ON THE BACK OF
THE ASSIGNMENT RECORDS THE GROUP'S COMMENTS. THE GROUP WILL BE
LOOKING TO SEE IF THE ASSIGNMENT IS CLEAR AND IF THE TWO EXAMPLES
REALLY SUPPORT THE POINT THE STUDENT IS MAKING. THE GROUP WILL
OFFER SUGGESTIONS FOR IMPROVEMENT AND NOTE WHAT THEY LIKED BEST.
THE PAPERS ALONG WITH THE GROUP COMMENTS ARE HANDED IN FOR
ASSESSMENT.

ASSIGNMENT # 39

GENERAL BIOLOGY I

LECTURE ANCHORING AND SUMMATION

TOPIC   MATTER AND ENERGY

THE ASSIGNMENT:

PART I

AT THE BEGINNING OF THE TOPIC ASK THE STUDENT TO DESCRIBE IN DETAIL THEIR DESK.

PART II

AT THE END OF THE TOPIC ASK THE STUDENT TO AGAIN DESCRIBE THE DESK BUT TO DO IT INCORPORATING THE INFORMATION ON ATOMIC STRUCTURE AND CHEMICAL BONDING JUST COVERED IN LECTURE.

THE IMPLEMENTATION:

PART I IS A 3-5 MINUTE FREE WRITING ASSIGNMENT DONE BEFORE I LECTURE ON ATOMIC STRUCTURE AND BONDING. THE STUDENT KEEPS THE ASSIGNMENT.

PART II IS A 3-5 MINUTE FREE WRITING ASSIGNMENT DONE AT THE END OF THE LECTURE TOPIC. AFTER THIS PART OF THE ASSIGNMENT IS DONE, PEER GROUPS ARE FORMED. EACH STUDENT READS BOTH PART I AND II TO THE GROUP AND THE GROUP SELECTS ONE SET TO PRESENT TO THE CLASS. AFTER THE GROUP PRESENTATIONS THERE IS A CLASS DISCUSSION ON WHAT IS REALITY.

AS THE GROUPS WORK I CIRCULATE AMONG THEM TO ASK QUESTIONS AND MAKE GENERAL AND SPECIFIC COMMENTS. I DO NOT COLLECT THE WORK.

ASSIGNMENT # 40

GENERAL BIOLOGY I

LECTURE ANCHORING

TOPIC   MAJOR COMPOUNDS

THE ASSIGNMENT:

Ask the students to free associate on fats, carbohydrates, and proteins.

THE IMPLEMENTATION:

Each student clusters on the topic for 2-3 minutes. The student then exchanges clusters with a neighbor and underlines two ideas or connections that are different from their own clusters. There is a class discussion as students volunteer ideas, words, and concepts from their clusters. I make a list of these ideas and words on the black board and incorporate them into the lecture material.

ASSIGNMENT # 41

GENERAL BIOLOGY I

LECTURE SUMMATION

TOPIC   MAJOR COMPOUNDS

THE ASSIGNMENT:

THE STUDENT IS ASKED TO COMPARE ANY TWO OF THE MAJOR CLASSES OF BIOLOGICAL COMPOUNDS.

THE IMPLEMENTATION:

THIS IS AN INDIVIDUAL 3-5 MINUTES FREE WRITING EXERCISE. AFTER THE 3-5 MINUTES PEER GROUPS ARE FORMED AND EACH STUDENT READS THEIR WORK TO THEIR GROUP. THE GROUPS THEN DISCUSS THEIR ANSWERS AND WRITE GROUP RESPONSES. THE GROUP RESPONSES ARE READ TO THE CLASS. USUALLY AT LEAST ONE GROUP WILL PRODUCE A LIST AND THIS LEADS INTO A DISCUSSION OF HOW TO WRITE A COMPARISON FOR AN ESSAY EXAM. I HAND OUT THE LIST OF EXAM WORDS AND WHAT THEY MEAN AND DISCUSS HOW TO WRITE ESSAY EXAMS QUESTIONS (SEE CHAPTER 10).

ASSIGNMENT # 42

GENERAL BIOLOGY I

LECTURE SUMMATION

TOPIC   EXAM QUESTIONS

THE ASSIGNMENT:

THE PURPOSE OF THIS ASSIGNMENT IS TO IMPROVE YOUR SUMMARIZATION SKILLS AS WELL AS TO GIVE YOU PRACTICE WRITING AND ANSWERING ESSAY QUESTIONS. AS YOU DO THE ASSIGNMENT YOU WILL BE ABLE TO IDENTIFY ANY SECTIONS OF THE MATERIAL WHERE YOU HAVE UNANSWERED QUESTIONS.

EACH STUDY GROUP IS TO WRITE AND ANSWER TWO SAMPLE ESSAY QUESTIONS.

THIS ASSIGNMENT WILL BE DONE BEFORE EACH EXAM.

THIS ASSIGNMENT IS DUE NEXT WEEK. EACH MEMBER OF THE STUDY GROUP IS TO SIGN THE ASSIGNMENT AS THE ENTIRE STUDY GROUP WILL RECEIVE THE SAME ASSESSMENT.

ASSESSMENT:

THE ASSIGNMENT WILL BE ASSESSED ON QUALITY OF THE QUESTION, COMPLETENESS AND CORRECTNESS OF THE ANSWER.

SCORING GUIDE:

3- DID THE TASK, QUALITY OF QUESTIONS, COMPLETE AND CORRECT ANSWERS, SHOWS SERIOUS EFFORT
2- DID THE TASK, QUALITY OF QUESTIONS, COMPLETENESS AND CORRECTNESS OF ANSWERS, WORK HASTY, BRIEF, SHOWS LITTLE EFFORT
1- CURSORY WORK
0- WORK NOT DONE

THE IMPLEMENTATION:

THE ASSIGNMENT IS HANDED OUT TWO WEEKS BEFORE THE EXAM AND IS DUE THE FOLLOWING WEEK. THIS ASSIGNMENT IS ASSIGNED BEFORE EACH OF THE FOUR EXAMS.
I READ AND MAKE COMMENTS ON THE ASSIGNMENTS AND ASSESS THEM ACCORDING TO THE ABOVE CRITERIA. I MAKE COPIES OF THE BEST QUESTIONS AND ANSWERS AND HAND THEM OUT IN CLASS. I BRIEFLY DISCUSS THE QUESTIONS WITH THE CLASS BEFORE THE EXAM.

ASSIGNMENT # 43

GENERAL BIOLOGY I

LECTURE ANCHORING

TOPIC   EXAM RESULTS

THE ASSIGNMENT:

PRESENT THE CLASS WITH A LIST OF THEIR UNORGANIZED TEST SCORES FROM THE FIRST EXAM. ASK THEM TO WORK WITH THE DATA AND PRESENT IT IN AN ORGANIZED FORM.

THE IMPLEMENTATION:

I PUT THE UNORGANIZED EXAM SCORES FROM THE FIRST EXAM ON THE BLACK BOARD AND ASK THE PEER GROUPS WHAT THE DATA MEANS TO THEM. I ASK THEM TO MAKE SOME SENSE OUT OF IT AND TO WRITE DOWN STEP BY STEP WHAT THEY DID TO MAKE SENSE OUT OF IT. EACH GROUP EXPLAINS TO THE CLASS WHAT THEY DID, WHY THEY DID IT, AND HOW THEY DID IT? THIS LEADS INTO A GENERAL DISCUSSION OF DATA MANIPULATION AND THE USE OF STATISTICS AND GRAPHIC REPRESENTATION.

COMMON STEPS FOR WORKING WITH DATA:

1- ORGANIZE DATA: HOW, WHY
2- CALCULATIONS: RANGE, AVERAGE, MEANS, NUMBER OF STUDENTS WHO RECEIVED EACH GRADE, PERCENTAGE FOR EACH GRADE, PERCENTAGE PASS/FAIL
3- GRAPHIC REPRESENTATION: TYPES WITH INFORMATIVE CAPTIONS
4- CONCLUSIONS.

ASSIGNMENT # 44

GENERAL BIOLOGY I

LECTURE ANCHORING

TOPIC   ENZYME

THE ASSIGNMENT:

DESIGN A EXPERIMENT TO DETERMINE THE OPTIMUM TEMPERATURE OF AN ENZYME.

THE IMPLEMENTATION:

THIS EXERCISE IS PART OF THE MATERIAL ON PROTEIN STRUCTURE AND FUNCTION. I GIVE AN EXAMPLE OF HOW TO DESIGN AN EXPERIMENT TO DETERMINE EXCESS SUBSTRATE CONCENTRATION AND THEN ASK THE CLASS TO FORM PEER GROUPS. EACH GROUP IS TO DESIGN AN EXPERIMENT TO DETERMINE OPTIMUM TEMPERATURE FOR AN ENZYME AND THEN GRAPH THE EXPECTED RESULTS. EACH GROUP WRITES AN EXPLANATION OF THEIR GRAPH BASED ON WHAT THEY KNOW ABOUT ENZYME STRUCTURE AND FUNCTION AND THE COLLISION THEORY. ALL MEMBERS OF THE GROUP SIGN THE WORK.

THERE IS A CLASS DISCUSSION OF THEIR WORK AND I DO A FINAL SUMMARY OF THE ENZYME TOPIC.

I COLLECT THEIR WORK, MAKE COMMENTS, AND RETURN IT AT THE NEXT LECTURE SESSION.

ASSIGNMENT # 45

GENERAL BIOLOGY I

LECTURE ANCHORING

TOPIC   CELL

THE ASSIGNMENT:

THE CLASS IS ASKED TO CLUSTER ON THE WORD "CELL".

THE IMPLEMENTATION:

EACH STUDENT CLUSTERS ON THE TOPIC FOR 2-3 MINUTES. THE STUDENT THEN EXCHANGES CLUSTERS WITH A NEIGHBOR AND UNDERLINES TWO IDEAS OR CONNECTIONS THAT ARE DIFFERENT FROM THEIR OWN CLUSTERS. THIS GIVES THE STUDENT THE EXPERIENCE OF SEEING THAT EVERYONE HAS A DIFFERENT EXPERIENTIAL BACKGROUND AND POINT OF VIEW. AS STUDENTS VOLUNTEER IDEAS, WORDS, AND CONCEPTS FROM THEIR CLUSTERS THERE DEVELOPES A DISCUSSION OF WHAT THE WORD CELL MEANS IN BIOLOGY AND IN OTHER DISCIPLINES. E.G. WHAT IS A PRISON CELL LIKE? HOW IS A PRISON CELL LIKE A BIOLOGICAL CELL? I MAKE A LIST OF THESE IDEAS AND WORDS ON THE BLACK BOARD AND INCORPORATE THEM INTO THE LECTURE MATERIAL.

ASSIGNMENT # 46

GENERAL BIOLOGY I

LECTURE SUMMATION

TOPIC   CELL MEMBRANE

THE ASSIGNMENT:

COMPARE THE DAVSON-DANIELLI AND THE SINGER-NICHOLSON MODELS OF MEMBRANE STRUCTURE.

THE IMPLEMENTATION:

THIS WRITING EXERCISE IS DONE INDIVIDUALLY AND THEN PEER GROUPS ARE FORMED. EACH PERSON READS THEIR COMPARISON TO THE GROUP AND THERE IS A DISCUSSION OF EACH WORK E.G. IS IT CLEAR, WHERE DOES IT NEED MORE DETAILED INFORMATION, IS THE QUESTION BEING ANSWERED. THE GROUP SELECTS ONE ESSAY TO READ TO THE CLASS. THIS EXERCISE GIVES THE STUDENT THE OPPORTUNITY TO SYNTHESIZE INFORMATION AS WELL AS PRACTICE WRITING ANSWERS TO ESSAY QUESTIONS. AFTER THE CLASS PRESENTATIONS THERE IS A DISCUSSION OF THE ESSAYS E.G. WHAT IS THE QUESTION ASKING FOR? WHAT INFORMATION IS NEEDED TO ANSWER THE QUESTION? HOW TO GO ABOUT IT.

AS THE GROUPS WORK I CIRCULATE AND PARTICIPATE OR MAKE COMMENTS IF NECESSARY.

ASSIGNMENT # 47

GENERAL BIOLOGY I

LECTURE SUMMATION

TOPIC   OSMOSIS

THE ASSIGNMENT:

YOU CATCH A TROUT (A FRESH WATER FISH) AND TAKE IT HOME TO
KEEP AS A PET BUT ALL YOU HAVE TO KEEP IT IN IS YOUR SALT WATER
FISH TANK. WHAT WOULD HAPPEN TO THE FISH AND WHY? EXPLAIN WHAT
WOULD HAPPEN TO THE FISH IN TERMS OF WATER AND SALT MOVEMENT.

THE IMPLEMENTATION:

THIS IS AN INDIVIDUAL EXERCISE FOLLOWED BY VOLUNTEER
READINGS. THE CLASS DISCUSSES WHAT HAPPENS TO THE FISH IN TERMS
OF OSMOSIS.
I COLLECT, READ, AND MAKE COMMENTS ON THESE PAPERS.

ASSIGNMENT # 48

GENERAL BIOLOGY I

LECTURE SUMMATION

TOPIC   ORGANELLES

THE ASSIGNMENT:

CHOOSE THE ONE ORGANELLE YOU LIKED THE BEST, DESCRIBE ITS STRUCTURE AND FUNCTION TO YOUR 10 YEAR OLD BROTHER, SISTER OR COUSIN AND EXPLAIN WHY IT IS YOUR FAVORITE. BE SURE TO USE WORDS OR IMAGES THAT YOUR LISTENER WILL UNDERSTAND AND RELATE TO. A TEN YEAR OLD IS IN THE FIFTH GRADE. USE YOUR IMAGINATION.

THE IMPLEMENTATION:

THIS IS DONE INDIVIDUALLY AND THEN PEER GROUPS ARE FORMED. EACH STUDENT READS THEIR ASSIGNMENT TO THE GROUP AND THE GROUP COMMENTS ON WHETHER THE EXPLANATION OF STRUCTURE AND FUNCTION WOULD BE CLEAR TO A TEN YEAR OLD OR NOT, DID THE STUDENT DO WHAT WAS ASKED. THE GROUP SELECTS ONE ASSIGNMENT AND READS IT TO THE CLASS. THIS LEADS INTO A SHORT DISCUSSION OF WAYS TO REMEMBER BY MAKING VIVID IMAGES OR MENTAL PICTURES E.G. A LYSOSOME IS LIKE A LARGE BALLOON WHICH HAVE LITTLE PAC MEN INSIDE, WHEN THE BALLOON BREAKS THE PAC MEN COME OUT AN EAT UP SPECIFIC THINGS.
I COLLECT THESE ASSIGNMENTS AND MAKE COMMENTS ON THEM.

ASSIGNMENT # 49

GENERAL BIOLOGY I

LECTURE ANCHORING

TOPIC   PHOTOSYNTHESIS

THE ASSIGNMENT:

THE CLASS IS ASKED TO CLUSTER ON THE WORD "LEAF".

THE IMPLEMENTATION:

EACH STUDENT CLUSTERS ON THE TOPIC FOR 2-3 MINUTES. THE
STUDENT EXCHANGES CLUSTERS WITH A NEIGHBOR AND UNDERLINES TWO
IDEAS OR CONNECTIONS THAT ARE DIFFERENT FROM THEIR OWN CLUSTERS.
THERE IS A CLASS DISCUSSION AS STUDENTS VOLUNTEER IDEAS, WORDS,
AND CONCEPTS FROM THEIR CLUSTERS. I MAKE A LIST OF THESE IDEAS
AND WORDS ON THE BLACK BOARD GROUPING THEM ACCORDING TO STRUCTURE
AND FUNCTION. I THEN INCORPORATE THIS LIST INTO THE LECTURE.

ASSIGNMENT # 50

GENERAL BIOLOGY I

LECTURE SUMMATION

TOPIC   PHOTOSYNTHESIS

THE ASSIGNMENT:

COMPARE CYCLIC AND NON-CYCLIC PHOTOPHOSPHORYLATION.

THE IMPLEMENTATION:

THE COMPARISON IS DONE INDIVIDUALLY. VOLUNTEERS THEN TO READ THEIR WORK TO THE CLASS AND I RECORD THEIR MAJOR POINTS ON THE BLACK BOARD. THERE IS CLASS DISCUSSION FOLLOWED BY A FINAL SYNTHESIS OF THE MATERIAL.

ASSIGNMENT # 51

GENERAL BIOLOGY I

LECTURE SUMMATION

TOPIC   PHOTOSYNTHESIS

THE ASSIGNMENT:

CONSTRUCT A LABELED DIAGRAM THAT DEMONSTRATES THE
RELATIONSHIPS (STRUCTURAL AND FUNCTIONAL) BETWEEN THE LIGHT AND
DARK REACTIONS OF PHOTOSYNTHESIS.

THE IMPLEMENTATION:

THIS EXERCISE IS DONE BY PEER GROUPS. EACH GROUP PUTS ITS
DIAGRAM ON THE BLACK BOARD. THERE IS A CLASS DISCUSSION OF EACH
DIAGRAM AND A SHORT DISCUSSION OF THE USE OF VISUAL
REPRESENTATION TO HELP TO ORGANIZE AND REMEMBER COMPLEX
MATERIAL.
I CIRCULATE AMONG THE GROUPS AS THEY WORK TO GIVE
ENCOURAGEMENT AND ANSWER QUESTIONS.

ASSIGNMENT # 52

GENERAL BIOLOGY I

LECTURE SUMMATION

TOPIC   PHOTOSYNTHESIS

THE ASSIGNMENT:

CLUSTER ON THE TERM "PHOTOSYNTHESIS".

THE IMPLEMENTATION:

1- THE ASSIGNMENT BEGINS WITH A CLASS DISCUSSION ON: HOW DO YOU DESIGN AN EXPERIMENT?

2- STUDENTS THEN INDIVIDUALLY CLUSTER ON THE TERM. THIS EXERCISE TELLS THEM WHAT THEY KNOW ABOUT THE TOPIC.

PARTS 3-6 ARE DONE IN PEER GROUPS.

3- THE GROUP CHOOSES ONE ITEM FROM THEIR CLUSTERS AND EACH STUDENT IN THE GROUP WRITES OUT THEY KNOW ABOUT THAT ITEM IN SENTENCE FORM.

4- THE GROUP CHOOSES ONE SENTENCE FROM THE INDIVIDUAL WORKS AND DECIDES HOW THIS WOULD BE TESTED IN THE LABORATORY.

5- THE GROUP DESIGNS AN EXPERIMENT INDICATING THE VARIABLES AND THE CONTROLS.

6- THE PEER GROUPS PRESENT THEIR EXPERIMENTS TO THE CLASS.

THE PRESENTATIONS ARE FOLLOWED BY CLASS DISCUSSION OF WHAT EACH GROUP HAS DONE. I RELATE THIS TO THEIR LABORATORY EXPERIENCES.

I CIRCULATE AMOUNT THE GROUPS AS THEY WORK TO GIVE ENCOURAGEMENT AND ANSWER QUESTIONS.

ASSIGNMENT # 53

GENERAL BIOLOGY I

LECTURE ANCHORING

TOPIC   CELLULAR RESPIRATION

THE ASSIGNMENT:

CLUSTER ON THE WORD "MITOCHONDRIA".

THE IMPLEMENTATION:

THIS IS AN INDIVIDUAL FREE ASSOCIATION EXERCISE TO FOCUS THE STUDENTS ON THE INFORMATION THEY ALREADY HAVE ABOUT THE MITOCHONDRIA. EACH STUDENT CLUSTERS ON THE TOPIC FOR 2-3 MINUTES. THE STUDENT THEN EXCHANGES CLUSTERS WITH A NEIGHBOR AND UNDERLINES TWO IDEAS OR CONNECTIONS THAT ARE DIFFERENT FROM THEIR OWN CLUSTERS. THERE IS A CLASS DISCUSSION AS STUDENTS VOLUNTEER IDEAS, WORDS, AND CONCEPTS FROM THEIR CLUSTERS. I MAKE A LIST OF THESE IDEAS AND WORDS ON THE BLACK BOARD AND INCORPORATE THEM INTO THE LECTURE ON THE STRUCTURE AND FUNCTION OF THE MITOCHONDRIA IN RELATIONSHIP TO CELLULAR RESPIRATION.

ASSIGNMENT # 54

GENERAL BIOLOGY I

LECTURE SUMMATION

TOPIC   CELLULAR RESPIRATION

THE ASSIGNMENT:

CONSTRUCT A LABELED DIAGRAM THAT DEMONSTRATES THE RELATIONSHIP BETWEEN THE VARIOUS METABOLIC PATHWAYS OF CELLULAR RESPIRATION AND THE STRUCTURE OF THE MITOCHONDRIA.

THE IMPLEMENTATION:

PEER GROUPS ARE FORMED AND DEVELOP DIAGRAMS. I CIRCULATE AMONG THE GROUPS AS THEY WORK TO ANSWER ANY QUESTIONS THEY MAY HAVE AND TO GIVE SUPPORTIVE COMMENTS. EACH GROUP PUTS THEIR DIAGRAM ON THE BLACK BOARD AND THEN THERE IS CLASS DISCUSSION. AFTER THE DISCUSSION I DO A FINAL SUMMARY OF CELLULAR RESPIRATION AND PUT UP A DIAGRAM OF MY OWN. I POINT OUT THE USEFULNESS OF USING DIAGRAMS AS A MEANS TO SYNTHESIZE INFORMATION, TO HELP LEARN NEW INFORMATION, AND AS A MEMORY AID.

ASSIGNMENT # 55

GENERAL BIOLOGY I

LECTURE ANCHORING

TOPIC   CELL REPRODUCTION

THE ASSIGNMENT:

CLUSTER ON THE TERM "CELL DIVISION".

THE IMPLEMENTATION:

THIS INDIVIDUAL FREE ASSOCIATION EXERCISE IS USED TO FOCUS THE STUDENTS ON THE INFORMATION THEY ALREADY HAVE ABOUT CELL DIVISION AND AS A LEAD INTO THE LECTURE ON CELL REPRODUCTION. EACH STUDENT CLUSTERS ON THE TOPIC FOR 2-3 MINUTES. THE STUDENT THEN EXCHANGES CLUSTERS WITH A NEIGHBOR AND UNDERLINES TWO IDEAS OR CONNECTIONS THAT ARE DIFFERENT FROM THEIR OWN CLUSTERS. THERE IS A CLASS DISCUSSION AS STUDENTS VOLUNTEER IDEAS, WORDS, AND CONCEPTS FROM THEIR CLUSTERS. I MAKE A LIST OF THESE IDEAS AND WORDS ON THE BLACK BOARD AND THE MATERIAL LEADS INTO THE LECTURE ON THE CELL CYCLE, MITOSIS, AND MEIOSIS.

ASSIGNMENT # 56

GENERAL BIOLOGY I

LECTURE SUMMATION

TOPIC   CELL REPRODUCTION

THE ASSIGNMENT:

COMPARE MITOSIS AND MEIOSIS.

THE IMPLEMENTATION:

AFTER THE EXERCISE IS DONE INDIVIDUALLY, PEER GROUPS ARE FORMED. EACH STUDENT READS THEIR COMPARISON TO THE GROUP. THE GROUP THEN DISCUSSES EACH WORK E.G. IS IT CLEAR, WHERE DOES IT NEED MORE DETAILED INFORMATION, IS THE QUESTION BEING ANSWERED. THE GROUP SELECTS ONE ESSAY AND READS IT TO TO THE CLASS. THIS EXERCISE GIVES THE STUDENT PRACTICE IN SYNTHESIZING INFORMATION AS WELL AS WRITING ANSWERS TO ESSAY QUESTIONS. THERE IS A CLASS DISCUSSION OF THE GROUPS' ESSAYS E.G. WHAT IS THE QUESTION ASKING FOR? WHAT INFORMATION IS NEEDED TO ANSWER THE QUESTION? HOW TO GO ABOUT IT.

AFTER THE CLASS DISCUSSION I GIVE THE FINAL SUMMARY OF THE LECTURE MATERIAL.

AS THE GROUPS WORK I CIRCULATE AND PARTICIPATE OR MAKE COMMENTS IF NECESSARY.

ASSIGNMENT # 57

GENERAL BIOLOGY I

LECTURE ANCHORING

TOPIC   NATURE OF THE GENE

THE ASSIGNMENT:

CLUSTER ON THE WORD "CHROMOSOME".

THE IMPLEMENTATION:

THIS AN INDIVIDUAL FREE ASSOCIATION EXERCISE USED TO FOCUS
THE STUDENTS ON THE INFORMATION THEY ALREADY HAVE ABOUT
CHROMOSOMES AND AS A LEAD INTO THE LECTURE ON THE NATURE OF THE
GENE AND GENE ACTION. EACH STUDENT CLUSTERS ON THE TOPIC FOR 2-3
MINUTES. THE STUDENT THEN EXCHANGES CLUSTERS WITH A NEIGHBOR AND
UNDERLINES TWO IDEAS OR CONNECTIONS THAT ARE DIFFERENT FROM THEIR
OWN CLUSTERS. THERE IS A CLASS DISCUSSION AS STUDENTS VOLUNTEER
IDEAS, WORDS, AND CONCEPTS FROM THEIR CLUSTERS. I MAKE A LIST OF
THESE IDEAS AND WORDS ON THE BLACK BOARD AND INCORPORATE THEM
INTO THE LECTURE ON THE STRUCTURE OF DNA.

ASSIGNMENT # 58

GENERAL BIOLOGY I

LECTURE SUMMATION

TOPIC   GENE ACTION

THE ASSIGNMENT:

CONSTRUCT A LABELED DIAGRAM THAT WILL EXPLAIN HOW THE INFORMATION STORED IN DNA (GENOTYPE) IS TRANSLATED INTO THE PHENOTYPE OF THE ORGANISM.

THE IMPLEMENTATION:

PEER GROUPS ARE FORMED AND DEVELOP DIAGRAMS. I CIRCULATE AMONG THE GROUPS AS THEY WORK TO ANSWER ANY QUESTIONS THEY MAY HAVE AND TO GIVE SUPPORTIVE COMMENTS. EACH GROUP PUTS THEIR DIAGRAM ON THE BLACK BOARD AND THEN THERE IS CLASS DISCUSSION. I DO A FINAL SUMMARY OF GENE ACTION AND PUT UP A DIAGRAM OF MY OWN. I ONCE MORE POINT OUT THE USEFULNESS OF USING DIAGRAMS TO SYNTHESIZE INFORMATION, TO HELP LEARN NEW INFORMATION AND AS A MEMORY AID.

ASSIGNMENT # 59

GENERAL BIOLOGY I

LECTURE ANCHORING

TOPIC   PATTERNS OF INHERITANCE AND GENETICS

THE ASSIGNMENT:

CLUSTER ON THE WORDS "MENDLE AND GENETICS ".

THE IMPLEMENTATION:

THIS INDIVIDUAL FREE ASSOCIATION EXERCISE IS USED TO FOCUS THE STUDENTS ON THE INFORMATION THEY ALREADY HAVE ABOUT MENDLE AND GENETICS AND AS A LEAD INTO THE LECTURE ON THE PATTERNS OF INHERITANCE AND GENETICS. EACH STUDENT CLUSTERS ON THE TOPIC FOR 2-3 MINUTES. THE STUDENT THEN EXCHANGES CLUSTERS WITH A NEIGHBOR AND UNDERLINES TWO IDEAS OR CONNECTIONS THAT ARE DIFFERENT FROM THEIR OWN CLUSTERS. THERE IS A CLASS DISCUSSION AS STUDENTS VOLUNTEER IDEAS, WORDS, AND CONCEPTS FROM THEIR CLUSTERS. I MAKE A LIST OF THESE IDEAS AND WORDS ON THE BLACK BOARD AND INCORPORATE THEM INTO THE LECTURE MATERIAL.

ASSIGNMENT # 60

GENERAL BIOLOGY I

LECTURE ANCHORING

TOPIC   GENETICS

THE ASSIGNMENT:

DO SEVERAL GENETICS PROBLEMS.

THE IMPLEMENTATION:

DURING THE LECTURE ON GENETICS I COVER A NUMBER OF DIFFERENT TYPES OF GENETIC PROBLEMS. I DO ONE OR MORE PROBLEMS OF EACH TYPE ON THE BLACK BOARD, EXPLAINING HOW THEY ARE DONE AND WHAT THEY CAN INDICATE ABOUT THE TRAITES BEING STUDIED. FOR EACH TYPE I THEN HAND OUT UNWORKED EXAMPLES AND HAVE THE STUDENTS INDIVIDUALLY WORK THEM. THE STUDENT THEN COMPARES THEIR WORK WITH A NEIGHBOR AND DISCUSSES ANY DISCREPANCIES. VOLUNTEERS PUT THEIR WORK ON THE BOARD FOR CLASS DISCUSSION. AS THE STUDENTS WORK I CIRCULATE AMONG THEM TO GIVE INDIVIDUAL ASSISTANCE.

ASSIGNMENT # 61

GENERAL BIOLOGY I

LECTURE ANCHORING

TOPIC   GENETICS

THE ASSIGNMENT:

CLUSTER ON THE WORD "MUTATION".

THE IMPLEMENTATION:

THIS INDIVIDUAL FREE ASSOCIATION EXERCISE IS USED TO FOCUS
THE STUDENTS ON THE INFORMATION THEY ALREADY HAVE ABOUT MUTATIONS
AND AS A LEAD INTO THE LECTURE ON MUTATIONS. EACH STUDENT
CLUSTERS ON THE TOPIC FOR 2-3 MINUTES. THE STUDENT THEN
EXCHANGES CLUSTERS WITH A NEIGHBOR AND UNDERLINES TWO IDEAS OR
CONNECTIONS THAT ARE DIFFERENT FROM THEIR OWN CLUSTERS. THERE IS
A CLASS DISCUSSION AS STUDENTS VOLUNTEER IDEAS, WORDS, AND
CONCEPTS FROM THEIR CLUSTERS. I MAKE A LIST OF THESE IDEAS AND
WORDS ON THE BLACK BOARD AND INCORPORATE THEM INTO THE LECTURE ON
MUTATIONS AND A DISCUSSION ON GENETIC COUNSELLING.

ASSIGNMENT # 62

GENERAL BIOLOGY II

LECTURE ANCHORING

TOPIC   LEVELS OF ORGANIZATION

THE ASSIGNMENT:

CLUSTER ON THE WORD "LEAF".

THE IMPLEMENTATION:

SINCE THIS IS THE FIRST TIME I HAVE USED CLUSTERING IN CLASS THIS SEMESTER, I FIRST EXPLAIN THE PROCESS AND GIVE AN EXAMPLE.
EACH STUDENT CLUSTERS ON THE TOPIC FOR 2-3 MINUTES. THE STUDENT THEN EXCHANGES CLUSTERS WITH A NEIGHBOR AND UNDERLINES TWO IDEAS OR CONNECTIONS THAT ARE DIFFERENT FROM THEIR OWN CLUSTERS.
BEFORE I CONTINUE WITH THE CLASS DISCUSSION ON THE LEVELS OF ORGANIZATION I DISCUSS THE USE OF CLUSTERING AS AN ALTERNATIVE TO PREPARING AN OUTLINE WHEN APPROACHING AN ESSAY EXAM QUESTION.
NEXT, I ASK THE CLASS TO GIVE ME THEIR IDEAS, WORDS, CONCEPTS, PHRASES OR CONNECTIONS AND I MAKE A LIST OF THESE ON THE BLACK BOARD. WHEN ALL OF THEIR IDEAS ARE ON THE BLACK BOARD I ARRANGE THEM INTO THE DIFFERENT LEVELS OF ORGANIZATION AND PROCEED TO THE LECTURE MATERIAL.
I DO NOT COLLECT THIS EXERCISE.

ASSIGNMENT # 63

GENERAL BIOLOGY II

LECTURE SUMMATION

TOPIC   LEVELS OF ORGANIZATION

THE ASSIGNMENT:

SUMMARIZE IN TWO OR THREE SENTENCES THE LECTURE MATERIAL ON THE LEVELS OF ORGANIZATION OF A PLANT.

THE IMPLEMENTATION:

THIS EXERCISE IS DONE INDIVIDUALLY AND THEN PEER GROUPS ARE FORMED. EACH STUDENT READS THEIR SUMMARY TO THE GROUP FOR COMMENTS. AFTER DISCUSSING EACH SUMMARY THE GROUP CAN EITHER CHOOSE ONE OR WRITE A GROUP SUMMARY TO READ TO THE CLASS.

AFTER THE GROUP PRESENTATIONS I LEAD THE CLASS IN A DISCUSSION OF WHAT A SUMMARY IS SUPPOSED TO DO AND POINT OUT HOW SUMMARIZING A LARGE BODY OF INFORMATION CAN GIVE THE STUDENT AN OVERVIEW THAT MAKES SENSE OUT OF ALL THE LITTLE PIECES. I POINT OUT THAT SEEING THE FOREST IS JUST AS IMPORTANT AS SEEING THE TREES.

I DO THE EXERCISE WITH THE CLASS SO THAT I CAN SHARE MINE WITH THEM.

I COLLECT THE SUMMARIES TO COMMENT ON AND RETURN THEM AT THE NEXT LECTURE SESSION.

ASSIGNMENT # 64

GENERAL BIOLOGY II

LECTURE ANCHORING

TOPIC   NUTRITION

THE ASSIGNMENT:

TO EAT A PIECE OF FOOD AND WRITE DOWN WHAT IS HAPPENING.

THE IMPLEMENTATION:

I BRING STICKS OF LICORICE OR ANY FOOD THAT HAS TO BE BITTEN INTO TO CLASS. I ASK EACH STUDENT TO WRITE DOWN STEP BY STEP WHAT HAPPENS AS THEY EAT THE FOOD. I THEN ASK THEM TO WATCH THEIR NEIGHBOR EAT AND WRITE DOWN ANYTHING NEW THEY OBSERVE. THE STUDENT EXCHANGES PAPERS WITH A NEIGHBOR AND UNDERLINES ANY OBSERVATIONS THAT ARE DIFFERENT FROM WHAT THEY RECORDED.

I ASK FOR VOLUNTEERS TO READ THEIR PAPERS TO THE CLASS AND THERE IS A CLASS DISCUSSION ABOUT THE EATING PROCESS. THIS LEADS INTO A DISCUSSION OF WHY WE EAT AND NUTRITION IN GENERAL.

ASSIGNMENT # 65

GENERAL BIOLOGY II

LECTURE ANCHORING

TOPIC   NUTRITION

THE ASSIGNMENT:

CLUSTER ON THE WORD "STOMACH."

THE IMPLEMENTATION:

EACH STUDENT CLUSTERS ON THE TOPIC FOR 2-3 MINUTES. THE STUDENT THEN EXCHANGES CLUSTERS WITH A NEIGHBOR AND UNDERLINES TWO IDEAS OR CONNECTIONS THAT ARE DIFFERENT FROM THEIR OWN CLUSTERS. THERE IS A CLASS DISCUSSION AS STUDENTS VOLUNTEER IDEAS, WORDS, AND CONCEPTS FROM THEIR CLUSTERS. I MAKE A LIST OF THESE IDEAS AND WORDS ON THE BLACK BOARD AND INCORPORATE THEM INTO THE LECTURE ON THE STRUCTURE AND FUNCTION OF THE STOMACH AS PART OF THE DIGESTIVE SYSTEM.

ASSIGNMENT # 66

GENERAL BIOLOGY II

LECTURE SUMMATION

TOPIC   NUTRITION

THE ASSIGNMENT:

CHOOSE THE PART OF THE DIGESTIVE SYSTEM YOU LIKE THE BEST, DESCRIBE ITS STRUCTURE AND FUNCTION TO YOUR 10 YEAR OLD BROTHER, SISTER OR COUSIN AND EXPLAIN WHY YOU LIKE IT THE BEST. BE SURE TO USE WORDS OR IMAGES THAT YOUR LISTENER WILL UNDERSTAND. A 10 YEAR OLD IS IN THE FIFTH GRADE. USE YOUR IMAGINATION.

THE IMPLEMENTATION:

THIS IS DONE INDIVIDUALLY AND THEN PEER GROUPS ARE FORMED AND EACH STUDENT READS THEIR ASSIGNMENT TO THE GROUP. THE GROUP COMMENTS ON WHETHER THE EXPLANATION OF STRUCTURE AND FUNCTION WOULD BE CLEAR TO A 10 YEAR OLD AND WHETHER THE STUDENT DID WHAT WAS ASKED. EACH GROUP SELECTS ONE PAPER TO READ TO THE CLASS. THIS LEADS INTO A SHORT DISCUSSION OF USING VIVID IMAGES OR MENTAL PICTURES TO REMEMBER.

I COLLECT, READ, MAKE COMMENTS ON, AND RETURN THESE PAPERS AT THE NEXT LECTURE SESSION.

ASSIGNMENT # 67

GENERAL BIOLOGY II

LECTURE SUMMATION

TOPIC   GAS EXCHANGE

THE ASSIGNMENT:

CHOOSE ONE HYPOTHESIS OF HOW STOMATES OPEN AND CLOSE AND EXPLAIN IT TO YOUR BEST FRIEND.

THE IMPLEMENTATION:

THIS IS AN INDIVIDUAL 5 MINUTE WRITING ASSIGNMENT. THE STUDENT CAN USE CLASS NOTES FOR REFERENCE. PEER GROUPS ARE FORMED AND EACH STUDENT READS THEIR WORK TO THE GROUP FOR COMMENTS. THE GROUP SHOULD ASK: IS THIS CLEAR, IS IT CORRECT, ARE THERE THINGS LEFT OUT. COMMENTS ARE RECORDED ON THE BACK OF THE PAPER. THE GROUP THEN SELECTS ONE PAPER OR WRITES A GROUP RESPONSE TO THE ASSIGNMENT TO READ TO THE CLASS. AFTER THE PRESENTATIONS THERE IS CLASS DISCUSSION FOLLOWED BY A FINAL SUMMARY OF THE MATERIAL.

I CIRCULATE AMONG THE GROUPS AS THEY WORK AND MAKE COMMENTS AND SUGGESTIONS. I MAKE SURE THAT I GET TO SEE WHAT EVERYONE HAS DONE.

ASSIGNMENT # 68

GENERAL BIOLOGY II

LECTURE ANCHORING

TOPIC   GAS EXCHANGE

THE ASSIGNMENT:

DESCRIBE HOW A FISH BREATHES I.E. WHAT DOES IT DO? WHAT DOES IT LOOK LIKE?

THE IMPLEMENTATION:

THIS IS AN INDIVIDUAL FIVE MINUTE WRITING ASSIGNMENT. THE STUDENT EXCHANGES PAPERS WITH A NEIGHBOR AND READS WHAT THEY HAVE WRITTEN. I THEN ASK FOR VOLUNTEERS TO READ TO THE CLASS AND I MAKE A LIST OF THEIR OBSERVATIONS ON THE BLACK BOARD. THERE IS A CLASS DISCUSSION ON POSSIBLE EXPLANATIONS FOR THE OBSERVATIONS WHICH LEADS INTO THE LECTURE MATERIAL.

ASSIGNMENT # 69

GENERAL BIOLOGY II

LECTURE ANCHORING

TOPIC   RESPIRATION

THE ASSIGNMENT:

SIT STILL AND BREATHE FOR A MINUTE AND THEN WRITE DOWN WHAT
HAPPENS TO YOUR BODY AS YOU BREATHE.

THE IMPLEMENTATION:

I ASK EACH STUDENT TO WRITE DOWN STEP BY STEP WHAT HAPPENS TO
THEM AS THEY BREATHE. I THEN ASK THEM TO WATCH THEIR NEIGHBOR
BREATHE AND TO WRITE DOWN ANYTHING NEW THEY OBSERVE. NEIGHBORS
EXCHANGE PAPERS AND UNDERLINE ANY OBSERVATION THAT IS DIFFERENT
FROM WHAT THEY HAVE WRITTEN.

I ASK FOR VOLUNTEERS TO READ THEIR PAPER TO THE CLASS AND
THIS LEADS INTO A DISCUSSION OF WHY WE BREATHE AND THE LECTURE
MATERIAL ON RESPIRATION.

ASSIGNMENT # 70

GENERAL BIOLOGY II

LECTURE SUMMATION

TOPIC   RESPIRATION

THE ASSIGNMENT:

SUMMARIZE THE RELATIONSHIP BETWEEN HEMOGLOBIN AND OXYGEN IN TERMS OF THE BINDING AND RELEASING OF OXYGEN IN THE SYSTEM.

THE IMPLEMENTATION:

THIS EXERCISE IS DONE INDIVIDUALLY AND THEN PEER GROUPS ARE FORMED. EACH STUDENT READS THEIR SUMMARY TO THE GROUP FOR COMMENTS. AFTER DISCUSSING EACH SUMMARY THE GROUP CAN EITHER CHOOSE ONE OR WRITE A GROUP SUMMARY TO PRESENT TO THE CLASS. AFTER THE PRESENTATIONS THERE IS CLASS DISCUSSION.

I DO THE EXERCISE WITH THE CLASS SO THAT I CAN SHARE MINE WITH THEM.

I COLLECT THIS EXERCISE AND MAKE COMMENTS ON THE PAPERS.

ASSIGNMENT # 71

GENERAL BIOLOGY II

LECTURE ANCHORING

TOPIC   KIDNEY

THE ASSIGNMENT:

CLUSTER ON THE WORD "KIDNEY".

THE IMPLEMENTATION:

EACH STUDENT CLUSTERS ON THE TOPIC FOR 2-3 MINUTES. THE
STUDENT THEN EXCHANGES CLUSTERS WITH A NEIGHBOR AND UNDERLINES
TWO IDEAS OR CONNECTIONS THAT ARE DIFFERENT FROM THEIR OWN
CLUSTERS. THERE IS A CLASS DISCUSSION AS STUDENTS VOLUNTEER
IDEAS, WORDS, AND CONCEPTS FROM THEIR CLUSTERS. I MAKE A LIST OF
THESE IDEAS AND WORDS ON THE BLACK BOARD AND INCORPORATE THEM
INTO THE LECTURE MATERIAL ON EXCRETION.

ASSIGNMENT # 72

GENERAL BIOLOGY II

LECTURE SUMMATION

TOPIC   KIDNEY

THE ASSIGNMENT:

TWO PATIENTS ENTER THE HOSPITAL WITH KIDNEY DISEASE. EACH PATIENT NEEDS TO BE PUT ON A KIDNEY MACHINE BUT THERE IS TIME AVAILABLE FOR ONLY ONE PATIENT. YOU ARE THE PERSON WHO WILL HAVE TO DECIDE WHICH PATIENT GETS TO USE THE MACHINE AND WHICH WILL NOT. WHAT INFORMATION WOULD HELP YOU TO MAKE THIS DECISION?

THE IMPLEMENTATION:

THIS IS AN INDIVIDUAL FIVE MINUTE WRITING ASSIGNMENT. THEN PEER GROUPS ARE FORMED AND EACH STUDENT READS THEIR WORK TO THE GROUP. THIS IS FOLLOWED BY A CLASS DISCUSSION ABOUT MAKING VALUE JUDGMENTS AND SITUATIONAL ETHICS.

THE CLASS THEN CHOOSES BIOGRAPHIES FOR THE TWO PATIENTS E.G. SEX, AGE, MARTIAL STATUS, FINANCIAL STATUS, AND WORKING HISTORY. EACH GROUP IS THEN ASKED TO DECIDE WHICH OF THE TWO PATIENTS WILL NOT RECEIVE TREATMENT AND TO WRITE A LETTER TO THE FAMILY OF THE PATIENT THAT WAS TURNED DOWN EXPLAINING THEIR REASONING. THE LETTERS ARE READ TO THE CLASS AND THERE IS FURTHER DISCUSSION.

I CIRCULATE AMONG THE GROUPS AS THEY WORK AND MAKE COMMENTS AND SUGGESTIONS. I MAKE SURE THAT I GET TO SEE WHAT EVERYONE HAS DONE.

I COLLECT THE GROUP LETTERS FOR COMMENTS.

ASSIGNMENT # 73

GENERAL BIOLOGY II

LECTURE ANCHORING

TOPIC   THERMOREGULATION

THE ASSIGNMENT:

DESCRIBE HOW AND EXPLAIN WHY A LIZARD OR A SNAKE ORIENTS
ITSELF IN RELATIONSHIP TO THE SUN EARLY IN THE MORNING AND IN THE
MIDDLE OF THE DAY.

THE IMPLEMENTATION:

THIS IS AN INDIVIDUAL FIVE MINUTE FREE WRITING EXERCISE. I
ASK FOR VOLUNTEERS TO READ WHAT THEY HAVE WRITTEN TO THE CLASS
AND THIS IS FOLLOWED BY CLASS DISCUSSION. THIS DISCUSSION LEADS
INTO THE LECTURE MATERIAL ON THERMOREGULATION.

ASSIGNMENT # 74

GENERAL BIOLOGY II

LECTURE ANCHORING

TOPIC   THERMOREGULATION

THE ASSIGNMENT:

DESCRIBE IN DETAIL WHAT HAPPENS TO YOU WHEN YOU GO OUTSIDE AND IT IS VERY COLD. HOW DOES YOUR BODY REACT? DESCRIBE IN DETAIL WHAT HAPPENS TO YOU WHEN YOU GET VERY HOT?

THE IMPLEMENTATION:

THIS IS AN INDIVIDUAL FIVE MINUTE FREE WRITING EXERCISE FOLLOWED BY CLASS DISCUSSION. I ASK FOR VOLUNTEERS TO READ WHAT THEY HAVE WRITTEN TO THE CLASS. AFTER A STUDENT HAS READ I ASK THE CLASS TO EXPLAIN WHAT IS HAPPENING AND WHY. THIS DISCUSSION LEADS INTO THE LECTURE MATERIAL ON THERMOREGULATION.

ASSIGNMENT # 75

GENERAL BIOLOGY II

LECTURE ANCHORING

TOPIC   TRANSPORT - PLANTS

THE ASSIGNMENT:

DESCRIBE IN DETAIL HOW A WILTED PLANT LOOKS AND WHAT HAPPENS
AFTER YOU WATER IT.

THE IMPLEMENTATION:

THIS IS AN INDIVIDUAL FIVE MINUTE FREE WRITING EXERCISE
FOLLOWED BY CLASS DISCUSSION. I ASK FOR VOLUNTEERS TO READ WHAT
THEY HAVE WRITTEN TO THE CLASS. AFTER THE STUDENTS HAVE READ I
ASK THE CLASS TO EXPLAIN WHAT IS HAPPENING AND WHY. THIS
DISCUSSION LEADS INTO THE LECTURE MATERIAL ON TRANSPORT IN
PLANTS.

ASSIGNMENT # 76

GENERAL BIOLOGY II

LECTURE SUMMATION

TOPIC   TRANSPORT - PLANTS

THE ASSIGNMENT:

USING THE TRANSPIRATION/COHESION/TENSION THEORY EXPLAIN HOW WATER GETS TO THE TOP OF A MAPLE TREE.

THE IMPLEMENTATION:

THE EXERCISE IS DONE INDIVIDUALLY AND THEN PEER GROUPS ARE FORMED. EACH STUDENT READS THEIR PAPER TO THE GROUP AND THERE IS A DISCUSSION OF EACH WORK E.G. IS IT CLEAR, WHERE DOES IT NEED MORE DETAILED INFORMATION, IS THE QUESTION BEING ANSWERED. THE GROUP SELECTS ONE ESSAY TO READ TO THE CLASS. THIS EXERCISE GIVES THE STUDENT THE OPPORTUNITY TO SYNTHESIZE INFORMATION AS WELL AS PRACTICE WRITING ANSWERS TO ESSAY QUESTIONS. THERE IS A CLASS DISCUSSION OF THE GROUPS' ESSAYS E.G. WHAT IS THE QUESTION ASKING FOR? WHAT INFORMATION IS NEEDED TO ANSWER THE QUESTION? HOW TO GO ABOUT IT.

AFTER THE CLASS DISCUSSION I GIVE THE FINAL SUMMARY OF THE LECTURE MATERIAL.

AS THE GROUPS WORK I CIRCULATE AND PARTICIPATE OR MAKE COMMENTS IF NECESSARY.

ASSIGNMENT # 77

GENERAL BIOLOGY II

LECTURE ANCHORING

TOPIC   TRANSPORT - HUMAN

THE ASSIGNMENT:

CLUSTER ON THE WORD "HEART".

THE IMPLEMENTATION:

EACH STUDENT CLUSTERS ON THE TOPIC FOR 2-3 MINUTES. THE STUDENT EXCHANGES CLUSTERS WITH A NEIGHBOR AND UNDERLINES TWO IDEAS OR CONNECTIONS THAT ARE DIFFERENT FROM THEIR OWN CLUSTERS. THERE IS A CLASS DISCUSSION AS STUDENTS VOLUNTEER IDEAS, WORDS, AND CONCEPTS FROM THEIR CLUSTERS. I MAKE A LIST OF THESE IDEAS AND WORDS ON THE BLACK BOARD AND INCORPORATE THEM INTO THE LECTURE MATERIAL ON THE HUMAN CIRCULATORY SYSTEM.

ASSIGNMENT # 78

GENERAL BIOLOGY II

LECTURE SUMMATION

TOPIC   TRANSPORT - BLOOD CLOT FORMATION

THE ASSIGNMENT:

CONSTRUCT A LABELED DIAGRAM TO DEMONSTRATE HOW A BLOOD CLOT IS FORMED.

THE IMPLEMENTATION:

PEER GROUPS ARE FORMED AND DEVELOP DIAGRAMS. I CIRCULATE AMONG THE GROUPS AS THEY WORK TO ANSWER ANY QUESTIONS THEY MAY HAVE AND TO GIVE SUPPORTIVE COMMENTS. EACH GROUP PUTS THEIR DIAGRAM ON THE BLACK BOARD FOR CLASS DISCUSSION. AFTER THE DISCUSSION I DO A FINAL SUMMARY OF HOW BLOOD CLOTS ARE FORMED USING A DIAGRAM OF MY OWN. I ONCE MORE POINT OUT THE USEFULNESS OF USING DIAGRAMS TO SYNTHESIZE INFORMATION, TO HELP LEARN NEW INFORMATION, AND AS A MEMORY AID.

ASSIGNMENT # 79

GENERAL BIOLOGY II

LECTURE SUMMATION

TOPIC   TRANSPORT

THE ASSIGNMENT:

COMPARE THE STRUCTURE AND FUNCTION OF THE HUMAN TRANSPORT SYSTEM WITH THOSE OF PLANTS.

THE IMPLEMENTATION:

THE EXERCISE IS DONE INDIVIDUALLY AND THEN PEER GROUPS ARE FORMED. EACH PERSON READS THEIR COMPARISON TO THE GROUP AND IT IS DISCUSSED E.G. IS IT CLEAR, WHERE DOES IT NEED MORE DETAILED INFORMATION, IS THE QUESTION BEING ANSWERED. THE GROUP SELECTS ONE ESSAY TO READ TO THE CLASS. THIS EXERCISE GIVES THE STUDENT THE OPPORTUNITY TO SYNTHESIZE INFORMATION AS WELL AS PRACTICE WRITING ANSWERS TO ESSAY QUESTIONS. THERE IS A CLASS DISCUSSION OF THE GROUPS' ESSAYS E.G. WHAT IS THE QUESTION ASKING FOR? WHAT INFORMATION IS NEEDED TO ANSWER THE QUESTION? HOW TO GO ABOUT IT.

AFTER THE CLASS DISCUSSION I GIVE THE FINAL SUMMARY OF THE LECTURE MATERIAL.

AS THE GROUPS WORK I CIRCULATE AND PARTICIPATE OR MAKE COMMENTS OR SUGGESTIONS.

ASSIGNMENT # 80

GENERAL BIOLOGY II

LECTURE ANCHORING

TOPIC   IMMUNITY

THE ASSIGNMENT:

CLUSTER ON THE WORD "IMMUNE".

THE IMPLEMENTATION:

EACH STUDENT CLUSTERS ON THE TOPIC FOR 2-3 MINUTES. THE
STUDENT EXCHANGES CLUSTERS WITH A NEIGHBOR AND UNDERLINES TWO
IDEAS OR CONNECTIONS THAT ARE DIFFERENT FROM THEIR OWN CLUSTERS.
THERE IS A CLASS DISCUSSION AS STUDENTS VOLUNTEER IDEAS, WORDS,
AND CONCEPTS FROM THEIR CLUSTERS. I MAKE A LIST OF THESE IDEAS
AND WORDS ON THE BLACK BOARD AND INCORPORATE THEM INTO THE
LECTURE MATERIAL ON THE HUMAN IMMUNE SYSTEM AND HOW IT WORKS.

ASSIGNMENT # 81

GENERAL BIOLOGY II

LECTURE SUMMATION

TOPIC   IMMUNE SYSTEM

THE ASSIGNMENT:

CONSTRUCT A LABELED DIAGRAM THAT WILL EXPLAIN HOW THE HUMAN IMMUNE SYSTEM WORKS.

THE IMPLEMENTATION:

PEER GROUPS ARE FORMED AND DEVELOP DIAGRAMS. I CIRCULATE AMONG THE GROUPS AS THEY WORK TO ANSWER ANY QUESTIONS THEY MAY HAVE AND TO GIVE SUPPORTIVE COMMENTS. EACH GROUP PUTS THEIR DIAGRAM ON THE BLACK BOARD FOR CLASS DISCUSSION. AFTER THE DISCUSSION I DO A FINAL SUMMARY OF THE HUMAN IMMUNE SYSTEM AND PUT UP A DIAGRAM OF MY OWN.

ASSIGNMENT # 82

GENERAL BIOLOGY II

LECTURE ANCHORING

TOPIC   CHEMICAL INTEGRATION

THE ASSIGNMENT:

DESCRIBE IN DETAIL WHAT HAPPENS TO A SUNFLOWER DURING COURSE OF A DAY.

THE IMPLEMENTATION:

THIS IS AN INDIVIDUAL 3-5 MINUTE FREE WRITING EXERCISE FOLLOWED BY CLASS DISCUSSION. I ASK FOR VOLUNTEERS TO READ WHAT THEY HAVE WRITTEN TO THE CLASS. AFTER THE STUDENT HAS READ I ASK THE CLASS TO EXPLAIN WHAT IS HAPPENING AND WHY. THIS DISCUSSION LEASD INTO THE LECTURE MATERIAL ON CHEMICAL INTEGRATION.

ASSIGNMENT # 83

GENERAL BIOLOGY II

LECTURE SUMMATION

TOPIC   CHEMICAL INTEGRATION

THE ASSIGNMENT:

PRETEND YOU ARE A WHEAT SEEDLING LYING ON TOP OF THE GROUND, DESCRIBE WHAT IS HAPPENING TO YOUR VARIOUS ORGANS AND EXPLAIN WHY IT IS HAPPENING.

THE IMPLEMENTATION:

THE EXERCISE IS DONE INDIVIDUALLY AND THEN PEER GROUPS ARE FORMED. EACH PERSON READS THEIR PAPER TO THE GROUP AND IT IS DISCUSSED E.G. IS IT CLEAR, WHERE DOES IT NEED MORE DETAILED INFORMATION, DOES THE WRITING DO WHAT WAS ASKED. THE GROUP SELECTS ONE ESSAY TO READ TO THE CLASS AND THERE IS A CLASS DISCUSSION E.G. WHAT IS THE QUESTION ASKING FOR? WHAT INFORMATION IS NEEDED TO ANSWER THE QUESTION? HOW TO GO ABOUT IT. THIS EXERCISE GIVES THE STUDENT PRACTICE WITH WORKING WITH THE LECTURE MATERIAL AND PUTTING IT INTO THEIR OWN WORDS.

AFTER THE CLASS DISCUSSION I GIVE THE FINAL SUMMARY OF THE LECTURE MATERIAL.

AS THE GROUPS WORK I CIRCULATE AND PARTICIPATE OR MAKE COMMENTS OR SUGGESTIONS.

ASSIGNMENT # 84

GENERAL BIOLOGY II

LECTURE ANCHORING

TOPIC   HORMONES

THE ASSIGNMENT:

CLUSTER ON THE WORD "HORMONE".

THE IMPLEMENTATION:

EACH STUDENT CLUSTERS ON THE TOPIC FOR 2-3 MINUTES. THE STUDENT EXCHANGES CLUSTERS WITH A NEIGHBOR AND UNDERLINES TWO IDEAS OR CONNECTIONS THAT ARE DIFFERENT FROM THEIR OWN CLUSTERS. THERE IS A CLASS DISCUSSION AS STUDENTS VOLUNTEER IDEAS, WORDS, AND CONCEPTS FROM THEIR CLUSTERS. I MAKE A LIST OF THESE IDEAS AND WORDS ON THE BLACK BOARD AND INCORPORATE THEM INTO THE LECTURE MATERIAL ON HORMONES IN HUMANS.

ASSIGNMENT # 85

GENERAL BIOLOGY II

LECTURE SUMMATION

TOPIC   HORMONES

THE ASSIGNMENT:

USING YOUR CLASS NOTES, CHOOSE ONE HORMONE AS AN EXAMPLE TO EXPLAIN HOW NEGATIVE FEEDBACK WORKS TO CONTROL ACTIVITY.

THE IMPLEMENTATION:

THE EXERCISE IS A FIVE MINUTE FREE WRITING DONE INDIVIDUALLY. PEER GROUPS ARE FORMED AND EACH STUDENT READS THEIR PAPER TO THE GROUP. THERE IS A DISCUSSION OF EACH STUDENT'S WORK E.G. IS IT CLEAR, WHERE DOES IT NEED MORE DETAILED INFORMATION, IS THE QUESTION BEING ANSWERED. THE GROUP SELECTS ONE ESSAY TO READ TO THE CLASS. THIS EXERCISE GIVES THE STUDENT THE OPPORTUNITY TO SYNTHESIZE INFORMATION AS WELL AS PRACTICE WRITING ANSWERS TO ESSAY QUESTIONS. FOLLOWING THE GROUP PRESENTATIONS THE CLASS DISCUSSES THE ESSAYS E.G. WHAT IS THE QUESTION ASKING FOR? WHAT INFORMATION IS NEEDED TO ANSWER THE QUESTION? HOW TO GO ABOUT IT.

AS THE GROUPS WORK I CIRCULATE AND PARTICIPATE OR MAKE COMMENTS IF NECESSARY.

I COLLECT THE PAPERS AND MAKE COMMENTS ON THEM.

ASSIGNMENT # 86

GENERAL BIOLOGY II

LECTURE SUMMATION

TOPIC   CHEMICAL INTEGRATION

THE ASSIGNMENT:

COMPARE PLANT AND ANIMAL CHEMICAL INTEGRATION.

THE IMPLEMENTATION:

THE EXERCISE IS DONE INDIVIDUALLY. PEER GROUPS ARE FORMED AND EACH STUDENT READS THEIR COMPARISON TO THE GROUP. THERE IS A DISCUSSION OF EACH STUDENT'S WORK E.G. IS IT CLEAR, WHERE DOES IT NEED MORE DETAILED INFORMATION, IS THE QUESTION BEING ANSWERED. EACH GROUP SELECTS ONE ESSAY TO READ TO THE CLASS OR WRITES A GROUP ESSAY. THIS EXERCISE GIVES THE STUDENT THE OPPORTUNITY TO SYNTHESIZE INFORMATION AS WELL AS PRACTICE WRITING ANSWERS TO ESSAY QUESTIONS. THE GROUP PRESENTATIONS ARE FOLLOWED BY DISCUSSION OF THE ESSAYS E.G. WHAT IS THE QUESTION ASKING FOR? WHAT INFORMATION IS NEEDED TO ANSWER THE QUESTION? HOW TO GO ABOUT IT.

AFTER THE CLASS DISCUSSION I GIVE THE FINAL SUMMARY OF THE LECTURE MATERIAL.

AS THE GROUPS WORK I CIRCULATE AND PARTICIPATE OR MAKE COMMENTS IF NECESSARY.

ASSIGNMENT # 87

GENERAL BIOLOGY II

LECTURE ANCHORING

TOPIC   NEURAL INTEGRATION

THE ASSIGNMENT:

CLUSTER ON THE WORD "NERVE".

THE IMPLEMENTATION:

EACH STUDENT CLUSTERS ON THE TOPIC FOR 2-3 MINUTES. THE
STUDENT EXCHANGES CLUSTERS WITH A NEIGHBOR AND UNDERLINES TWO
IDEAS OR CONNECTIONS THAT ARE DIFFERENT FROM THEIR OWN CLUSTERS.
THERE IS A CLASS DISCUSSION AS STUDENTS VOLUNTEER IDEAS, WORDS,
AND CONCEPTS FROM THEIR CLUSTERS. I MAKE A LIST OF THESE IDEAS
AND WORDS ON THE BLACK BOARD AND INCORPORATE THEM INTO THE
LECTURE MATERIAL ON NEURAL INTEGRATION.

ASSIGNMENT # 88

GENERAL BIOLOGY II

LECTURE SUMMATION

TOPIC   NEURAL INTEGRATION

THE ASSIGNMENT:

CONSTRUCT A LABELED DIAGRAM WHICH WILL EXPLAIN THE STRUCTURE OF A NERVE AND HOW IT FUNCTIONS.

THE IMPLEMENTATION:

THE EXERCISE IS DONE INDIVIDUALLY AND THEN PEER GROUPS ARE FORMED. EACH STUDENT EXPLAINS THEIR DIAGRAM TO THE GROUP AND IT IS DISCUSSED E.G. IS IT CLEAR, WHERE DOES IT NEED MORE DETAILED INFORMATION, DOES THE DIAGRAM ANSWER THE QUESTION BEING ASKED. THE GROUP SELECTS ONE DIAGRAM OR CONSTRUCTS A GROUP DIAGRAM WHICH IS TO BE PUT ON THE BLACK BOARD FOR CLASS DISCUSSION. THIS EXERCISE GIVES THE STUDENT PRACTICE IN SYNTHESIZING INFORMATION AND PRESENTING IT IN A VISUAL FORM.

AS THE GROUPS WORK I CIRCULATE AND ASK QUESTIONS ABOUT THE DIAGRAMS.

ASSIGNMENT # 89

GENERAL BIOLOGY II

LECTURE ANCHORING

TOPIC   MOTILITY

THE ASSIGNMENT:

CLUSTER ON THE WORD "MUSCLE".

THE IMPLEMENTATION:

EACH STUDENT CLUSTERS ON THE TOPIC FOR 2-3 MINUTES. THE STUDENT EXCHANGES CLUSTERS WITH A NEIGHBOR AND UNDERLINES TWO IDEAS OR CONNECTIONS THAT ARE DIFFERENT FROM THEIR OWN CLUSTERS. THERE IS A CLASS DISCUSSION AS STUDENTS VOLUNTEER IDEAS, WORDS, AND CONCEPTS FROM THEIR CLUSTERS. I MAKE A LIST OF THESE IDEAS AND WORDS ON THE BLACK BOARD AND INCORPORATE THEM INTO THE LECTURE MATERIAL ON MOTILITY.

ASSIGNMENT # 90

GENERAL BIOLOGY II

LECTURE SUMMATION

TOPIC   MOTILITY

THE ASSIGNMENT:

DISCUSS THE STRUCTURE AND CONTROL OF MUSCLE CONTRACTION.

THE IMPLEMENTATION:

THE EXERCISE IS A FIVE MINUTE FREE WRITING DONE INDIVIDUALLY AND THEN PEER GROUPS ARE FORMED. EACH PERSON READS THEIR PAPER TO THE GROUP AND IT IS DISCUSSED E.G. IS IT CLEAR, WHERE DOES IT NEED MORE DETAILED INFORMATION, IS THE QUESTION BEING ANSWERED. THE GROUP SELECTS ONE ESSAY TO READ TO THE CLASS. THIS EXERCISE GIVES THE STUDENT THE OPPORTUNITY TO SYNTHESIZE INFORMATION AS WELL AS PRACTICE WRITING ANSWERS TO ESSAY QUESTIONS. THERE IS A CLASS DISCUSSION OF THE GROUP ESSAYS E.G. WHAT IS THE QUESTION ASKING FOR? WHAT INFORMATION IS NEEDED TO ANSWER THE QUESTION? HOW TO GO ABOUT IT.

AS THE GROUPS WORK I CIRCULATE AND PARTICIPATE OR MAKE COMMENTS IF NECESSARY.

AFTER THE CLASS DISCUSSION I GIVE A FINAL SUMMARY OF THE MATERIAL.

I COLLECT THE PAPERS AND MAKE COMMENTS ON THEM.

ASSIGNMENT # 91

GENERAL BIOLOGY II

LECTURE SUMMATION

TOPIC   MOTILITY

THE ASSIGNMENT:

CONSTRUCT A LABELED DIAGRAM WHICH WILL EXPLAIN HOW THE INFORMATION RECEIVED BY A RECEPTOR RESULTS IN THE ACTION OF AN EFFECTOR.

THE IMPLEMENTATION:

THE EXERCISE IS DONE INDIVIDUALLY AND THEN PEER GROUPS ARE FORMED. EACH STUDENT EXPLAINS THEIR DIAGRAM TO THE GROUP AND IT IS DISCUSSED E.G. IS IT CLEAR, WHERE DOES IT NEED MORE DETAILED INFORMATION, DOES THE DIAGRAM ANSWER THE QUESTION BEING ASKED. THE GROUP SELECTS ONE DIAGRAM OR CONSTRUCTS A GROUP DIAGRAM TO PUT ON THE BLACK BOARD FOR CLASS DISCUSSION. THIS EXERCISE GIVES THE STUDENT PRACTICE IN SYNTHESIZING INFORMATION AND PRESENTING IT IN A VISUAL FORM. THERE IS A CLASS DISCUSSION OF THE GROUPS' DIAGRAMS.

AS THE GROUPS WORK I CIRCULATE AND ASK QUESTIONS.

AFTER THE CLASS DISCUSSION I DO A FINAL SUMMARY OF THE TOPIC AND PUT MY OWN DIAGRAM ON THE BOARD.

ASSIGNMENT # 92

GENERAL BIOLOGY II

LECTURE ANCHORING

TOPIC   EVOLUTION

THE ASSIGNMENT:

CLUSTER ON THE WORD "EVOLUTION".

THE IMPLEMENTATION:

EACH STUDENT CLUSTERS ON THE TOPIC FOR 2-3 MINUTES. THE STUDENT EXCHANGES CLUSTERS WITH A NEIGHBOR AND UNDERLINES TWO IDEAS OR CONNECTIONS THAT ARE DIFFERENT FROM THEIR OWN CLUSTERS. THERE IS A CLASS DISCUSSION AS STUDENTS VOLUNTEER IDEAS, WORDS, AND CONCEPTS FROM THEIR CLUSTERS. I MAKE A LIST OF THESE IDEAS AND WORDS ON THE BLACK BOARD AND INCORPORATE THEM INTO THE LECTURE MATERIAL ON EVOLUTION.

ASSIGNMENT # 93

GENERAL BIOLOGY II

LECTURE SUMMATION

TOPIC   EVOLUTION

THE ASSIGNMENT:

DISCUSS THE CONCEPT OF SPECIATION.

THE IMPLEMENTATION:

THE EXERCISE IS A FIVE MINUTE WRITING DONE INDIVIDUALLY AND
THEN PEER GROUPS ARE FORMED. EACH PERSON READS THEIR PAPER TO
THE GROUP AND IT IS DISCUSSED E.G. IS IT CLEAR, WHERE DOES IT
NEED MORE DETAILED INFORMATION, IS THE QUESTION BEING ANSWERED.
THE GROUP SELECTS ONE ESSAY TO READ TO THE CLASS. THIS EXERCISE
GIVES THE STUDENT THE OPPORTUNITY TO SYNTHESIZE INFORMATION AS
WELL AS PRACTICE WRITING ANSWERS TO ESSAY QUESTIONS. AFTER
THE GROUP PRESENTATIONS THERE IS A CLASS DISCUSSION E.G. WHAT IS
THE QUESTION ASKING FOR? WHAT INFORMATION IS NEEDED TO ANSWER THE
QUESTION? HOW TO GO ABOUT IT.
AS THE GROUPS WORK I CIRCULATE AND PARTICIPATE OR MAKE
COMMENTS IF NECESSARY.
I COLLECT THE PAPERS AND MAKE COMMENTS ON THEM.

ASSIGNMENT # 94

GENERAL BIOLOGY II

LECTURE ANCHORING

TOPIC   ECOLOGY

THE ASSIGNMENT:

CLUSTER ON THE WORD "ECOLOGY".

THE IMPLEMENTATION:

EACH STUDENT CLUSTERS ON THE TOPIC FOR 2-3 MINUTES. THE
STUDENT EXCHANGES CLUSTERS WITH A NEIGHBOR AND UNDERLINES TWO
IDEAS OR CONNECTIONS THAT ARE DIFFERENT FROM THEIR OWN CLUSTERS.
THERE IS A CLASS DISCUSSION AS STUDENTS VOLUNTEER IDEAS, WORDS,
AND CONCEPTS FROM THEIR CLUSTERS. I MAKE A LIST OF THESE IDEAS
AND WORDS ON THE BLACK BOARD AND INCORPORATE THEM INTO THE
LECTURE MATERIAL ON ECOLOGY.

ASSIGNMENT # 95

GENERAL BIOLOGY II

LECTURE SUMMATION

TOPIC   ECOLOGY

THE ASSIGNMENT:

COMPARE THE THREE KINDS OF ECOLOGICAL PYRAMIDS AND DISCUSS THEIR LIMITATIONS.

THE IMPLEMENTATION:

THE EXERCISE IS DONE INDIVIDUALLY AND THEN PEER GROUPS ARE FORMED. EACH PERSON READS THEIR COMPARISON TO THE GROUP AND IT IS DISCUSSED E.G. IS IT CLEAR, WHERE DOES IT NEED MORE DETAILED INFORMATION, IS THE QUESTION BEING ANSWERED. THE GROUP SELECTS ONE ESSAY TO READ TO THE CLASS OR WRITES A GROUP RESPONSE. THIS EXERCISE GIVES THE STUDENT THE OPPORTUNITY TO SYNTHESIZE 'IFORMATION AS WELL AS PRACTICE WRITING ANSWERS TO ESSAY ESTIONS. AFTER THE GROUP PRESENTATIONS THERE IS A CLASS CUSSION E.G. WHAT IS THE QUESTION ASKING FOR? WHAT .FORMATION IS NEEDED TO ANSWER THE QUESTION? HOW TO GO ABOUT IT. THE DISCUSSION IS BROADENED TO INCLUDE THE IMPLICATIONS THAT THE ECOLOGICAL PYRAMIDS HAVE FOR THE WORLD TODAY.

AFTER THE CLASS DISCUSSION I GIVE THE FINAL SUMMARY OF THE LECTURE MATERIAL.

AS THE GROUPS WORK I CIRCULATE AND MAKE COMMENTS OR SUGGESTIONS.